POWER SYSTEMS CONTROL AND RELIABILITY

Electric Power Design and Enhancement

POWER SYSTEMS CONTROL AND RELIABILITY

Electric Power Design and Enhancement

POWER SYSTEMS CONTROL AND RELIABILITY

Electric Power Design and Enhancement

Isa S. Qamber

AAP | APPLE ACADEMIC PRESS

Apple Academic Press Inc.
4164 Lakeshore Road
Burlington ON L7L 1A4
Canada

Apple Academic Press Inc.
1265 Goldenrod Circle NE
Palm Bay, Florida 32905
USA

International Standard Book Number-13: 978-1-77188-821-9 (Hardcover)
International Standard Book Number-13: 978-0-42928-701-5 (eBook)

Library and Archives Canada Cataloguing in Publication

Title: Power systems control and reliability : electric power design and enhancement / Isa S. Qamber.

Names: Qamber, Isa S., author.

Description: Includes bibliographical references and index.

Identifiers: Canadiana (print) 20190239840 | Canadiana (ebook) 20190239875 | ISBN 9781771888219 (hardcover) | ISBN 9780429287015 (ebook)

Subjects: LCSH: Electric power systems. | LCSH: Reliability (Engineering)

Classification: LCC TK1001 .Q36 2020 | DDC 621.31—dc23

CIP data on file with US Library of Congress

Apple Academic Press also publishes its books in a variety of electronic formats. Some content that appears in print may not be available in electronic format. For information about Apple Academic Press products, visit our website at **www.appleacademicpress.com** and the CRC Press website at **www.crcpress.com**

About the Author

Isa S. Qamber, PhD

Professor, Electrical and Electronics Engineering Department, University of Bahrain, Kingdom of Bahrain

Isa S. Qamber, PhD, is a Professor in the Electrical and Electronics Engineering Department at the University of Bahrain. He joined the Department of Electrical Engineering and Computer Science, College of Engineering, University of Bahrain, as an assistant professor and was an ex-chairman for the Department of Electrical and Electronics Engineering in the same university. From 2008 to 2011, he was the Dean of the College of Applied Studies, University of Bahrain, and Dean of Scientific Research from 2011 to 2014. He is a senior member of the IEEE (Institute of Electrical and Electronics Engineers, USA) and was a member of the International Council on Large Electric Systems, the Bahrain Society of Engineers, and the Board of Bahrain Society of Engineers. He established and chaired the IEEE Bahrain Section. His research interests include studying the reliability of power systems and plants as well as renewable energy. Dr. Qamber received his BSc degree in electrical engineering from King Saud University, Kingdom of Saudi Arabia; his MSc degree from the University of Manchester Institute of Science and Technology, United Kingdom; and his PhD degree in reliability engineering from the University of Bradford, UK.

Contents

Contents

Abbreviations

ANFIS	adaptive network-based fuzzy inference system
ANN	artificial neural networks
APE	absolute percent error
ARK2	additive Runge-Kutta
CAIDI	customer average interruption duration index
CRM	capacity reserve margin
EENS	expected energy not served
EIR	energy index of reliability
EPLF	electric power load forecasting
EPNS	expected power not supplied
ERK	explicit Runge-Kutta
ESDIRK	explicit, singly-diagonally implicit Runge-Kutta
EUE	expected un-served energy
FI	fuzzy inference
FIS	fuzzy inference system
FL	fuzzy logic
FOR	forced outage rate
HLI	hierarchical level one
HLII	hierarchical level two
HLIII	hierarchical level three
HLOLE	hourly loss of load expectation
IDI	interruption duration index
LOEP	loss of energy probability
LOLE	loss of load expectation
LOLEV	loss of load events
LOLH	loss of load hours
LOLP	loss of load probability
LTLF	long-term load forecasting
MDT	mean downtime
MLP	multilayer perceptions
MTBF	mean time between fail
MTLF	medium-term load forecast
MTTF	mean time to fail

MTTR	mean time to repair
NERC	North American Electric Reliability Corporation
ORR	outage replacement rate
PLM	priority list method
PRM	planning reserve margin
RBD	reliability block diagram
SMM	system multiplication method
STLF	short term load forecast
VSTLF	very short-term load forecast

Preface

The subject of this book is the power systems reliability and generating unit commitments. This book presents some applications of probability methods and concepts in the field of electric power systems reliability. Many reliability applications used in the book will help students, engineers, and researchers in real life.

The present book will help readers who are familiar or unfamiliar with basic reliability concepts. These concepts help with the basics of reliability engineering, which are used in different fields of engineering. Furthermore, the electric power systems examples are illustrated in the book. The text contains educational (academic) and research ideas for the readers. Some applications are shown through a number of examples. The text is geared for undergraduate and graduate students as well as researchers. References used are included at the end of each chapter, which readers can refer to when they require further information and ideas.

The probability theory placed an important role in reliability applications. The probability application in the electric power systems led to the development of mathematical models illustrated in the present book based on need.

The reliability techniques help with evaluating the electric power systems for future design, planning, control, and operation. Basic appropriate and suitable algorithms are presented throughout the chapters. These algorithms help the researchers to implement their own suitable programs where needed.

The first four chapters contain the basic principles for the reliability concepts. Chapter 1 covers the unit commitment, the spinning reserve, and reserve markets, while Chapter 2 deals with the electric load curves, their benefits, and their objectives. Chapter 3 presents the power systems operation with two possibilities, while not taking into consideration the losses on the system, and on the other hand, while taking into consideration the losses. The power interchange is discussed and explained by applying the application of two methods. These methods are the priority list method and the unit-commitment solution method. The basic power systems reliability is also defined in the same chapter. A number of state models are explained and listed in Chapter 4 with important and needed indices.

Chapters 5 and 6 are devoted to the individual and cumulative probabilities and basic definitions of indices. In addition, components with different connections and different distributions are presented along with important indices. Moreover, the energy curtailed is calculated, where the PJM methods are presented with the frequency and duration method followed by the minimal cut-set method with a solved bridge network example. A number of developed models can be found in Chapter 6.

The Kronecker technique is discussed in Chapter 7, with a number of applications and examples explained, showing how to build power plants. The Kronecker technique is compared with the differentiation technique. By forming the plant(s), the transient probabilities that any model is passing through can be calculated using a suitable method presented in Chapter 8. These methods are the fourth-order Runge-Kutta, system multiplication method, or Adams method with different incremental times. Chapter 9 introduces the load forecasting, fuzzy logic, neuro-fuzzy systems, and fuzzy inference system. The estimated load forecast for the Kingdom of Bahrain is calculated by finding the model based on the actual data of the country. Chapter 10 deals with the equivalent transition rate matrix and using the Laplace transforms technique to calculate the transient probabilities that the model is passing through for a number of years. In addition, the percentage error of the peak loads is calculated and plotted for more than one model. Finally, a comparison between different state models compared.

Finally, I would like to thank my wife, Somaya Aljowder, for her constant encouragement to me in preparing this book. At the same time, I would like to thank my son, Abdulla, and my daughters, Aysha and Ameena, for their help during the preparation of the manuscript and their assistance in typing and preparing the figures.

—Isa S. Qamber, PhD

CHAPTER 1

Introduction and Literature Review

1.1 BASIC TERMS

In the reliability studies of power systems, adequacy and operating reliability are the two terms used. The first term, adequacy, is defined as the ability of the electricity system to supply the aggregate electric power demand and energy requirements of the end-user customers at any instants. It is needed to take into consideration the scheduled and reasonably expected unscheduled outages of system elements. The other term needs to be taking, as well; in consideration is the operating reliability, which is the ability of the bulk-power system to withstand sudden disturbances. Some of the disturbances are the electric short circuits or the unanticipated loss of system elements from credible contingencies while avoiding damage to equipment or uncontrolled cascading blackouts.

Regarding adequacy, the system operators should take controlled actions or procedures in order to keep them in a specified state and to have a continual balance between supply and demand by maintaining load resource balance within a specific area. At the same time, the system adequacy relates to an exist sufficient generators within the existed system to satisfy the consumer load demand (Boroujeni et al., 2012).

1.2 UNIT COMMITMENT

The unit commitment (Saravanan et al., 2016) is one of the decision-making problems in the electric power system. It helps to attain the main objective of minimizing the cost of operation of the generator by selecting the combination or group of ON/OFF generator after meeting all the constraints. A question raised to the researcher, *what is the meaning of the term commit?*

As a simple definition of that, the generators will share the loads (Saravanan et al., 2014). Therefore, in this case, the unit(s) will be in

ON mode. At the same time, the unit will help in satisfying the required load. Figure 1.1 shows a simple power network (Ananda et al., 2018; Jayabarathia and Jismab, 2015; Wang et al., 2018; Zhanga et al., 2019).

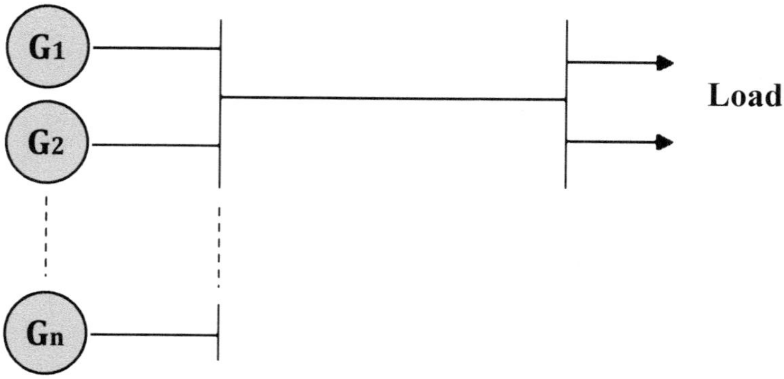

FIGURE 1.1 Simple power network.

1.3 SPINNING RESERVE

As a simple model to explain the spinning reserve is represented in Figure 1.2. As a simple model, it is represented by two generators which shows the generation side of the model, where the transmission of the power will be within the transmission line (the power loss is considered) and the model ended by a load point (Hreinsson et al., 2015; Palacio, 2015; Wang et al., 2017).

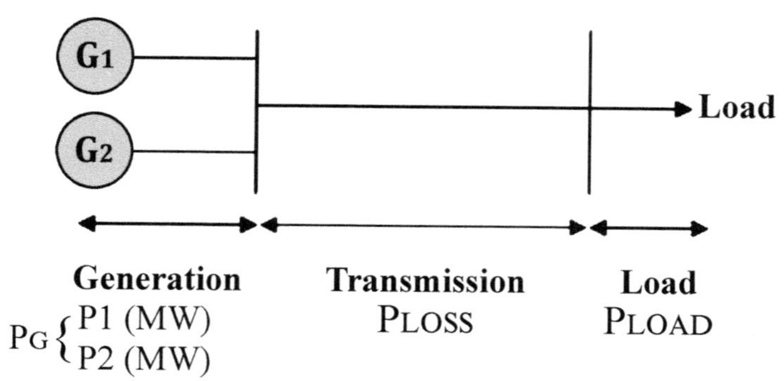

FIGURE 1.2 Simple electric power model (P_G: power generation; P_{LOSS}: power loss; P_{LOAD}: power load).

Spinning reserve depends on three factors:

1. P_G;
2. P_{LOSS}; and
3. P_{LOAD}.

The formula which combines the above factors with relation to the spinning reserve is written as:

$$P_G = P_{LOSS} + P_{LOAD} \tag{1}$$

$$P_G - P_{LOSS} - P_{LOAD} = 0 \tag{2}$$

$$P_G - P_{LOSS} - P_{LOAD} = \Delta P \tag{3}$$

where ΔP is the spinning reserve. Equation (2) shows the case of no spinning reserve exists, where Equation (3) has a difference in power ΔP called spinning reserve.

1.4 RESERVE MARKETS

Reserve markets might be defined as the electricity reserves playing a key role in stabilizing the power system (Allen and Ilic, 2000). This means that the market frequency remains within a stable band. In some countries, the reserves are divided between the cities or between countries in case of interconnection between them. For instance, in Finland, the Finnish TSO, Fingrid (Haakana et al., 2017), arranges the reserve market. A challenge in the reserve market participation showing that the volume of electricity reserve markets is limited. This gives a meaning that new market participants may have a considerable effect on the reserve markets. The reserve market will help in decreasing the reserve prices, too (Bergha et al., 2018; Ghahary and Abdollahi, 2018; Pandžića et al., 2018). A point has to be pointed that the reserve market participation may cause fluctuations in the heat delivery, and even minor shortages in heat delivery may occur in case that the up-regulated the electrical power for a long time (Haakana et al., 2017). One of the important matters to be involved in reserve markets is to protect the user producer against failure, where the producer may purchase reserve electric power from other power producers. The power traded in advance of its use, i.e., scheduled (Allen et al., 2000).

KEYWORDS

- **reserve markets**
- **spinning reserve**
- **unit commitment**

REFERENCES

Allen, E. H., & Ilic, M. D. Reserve markets for power systems reliability. *IEEE Transactions on Power Systems*, **2000**, *15*(1), 228–233.

Ananda, H., Naranga, N., & Dhillon, J. S. Unit commitment considering dual-mode combined heat and power generating units using integrated optimization technique. *Energy Conversion and Management*, **2018**, *171*, 984–1001.

Bergha, K. V., Bruninxa, K., & Delarue, E. Cross-border reserve markets: Network constraints in cross-border reserve procurement. *Energy Policy*, **2018**, *113*, 193–205.

Boroujeni, H. F., Eghtedari, M., Abdollahi, M., & Behzadipour, E. Calculation of generation system reliability index: Loss of load probability. *Life Science Journal*, **2012**, *9*(4), 4903–4908.

Ghahary, K., Abdollahi, A., Rashidineja, M., & Alizadeh, M. I. Optimal reserve market clearing considering uncertain demand response using information gap decision theory. *Electrical Power and Energy Systems*, **2018**, *101*, 213–222.

Haakana, J., Tikka, V., Lassila, J., & Partanen, J. Methodology to analyze combined heat and power plant operation considering electricity reserve market opportunities. *International Journal of Energy*, **2017**, *127*, 408–418.

Hreinsson, K., Vrakopoulou, M., & Andersson, G. Stochastic security constrained unit commitment and non-spinning reserve allocation with performance guarantees. *Electrical Power and Energy Systems*, **2015**, *72*, 109–115.

Jayabarathia, R., Jismab, M., & Suyampulingam, A. Unit commitment using embedded systems. *Procedia Technology*, **2015**, *21*, 96–102.

Palacio, S. N., Kircher, K. J., & Zhang, K. M. On the feasibility of providing power system spinning reserves from thermal storage. *Energy and Buildings*, **2015**, *104*, 131–138.

Pandžića, H., Dvorkinb, Y., & Carrión, M. Investments in merchant energy storage: Trading-off between energy and reserve markets. *Applied Energy*, **2018**, *230*, 277–286.

Saravanan, B., Kumar, C., & Kothari, D. P. A solution to unit commitment problem using fireworks algorithm. *Electrical Power and Energy Systems*, **2016**, *77*, 221–227.

Saravanan, B., Vasudevan, E. R., & Kothari, D. P. Unit commitment problem solution using invasive weed optimization algorithm. *Electrical Power and Energy Systems*, **2014**, *55*, 21–28.

Wang, J., Guo, M., & Liu, Y. Hydropower unit commitment with nonlinearity decoupled from mixed integer nonlinear problem. *Energy*, **2018**, *150*, 839–846.

Wang, S., Hui, H., Ding, Y., & Zhu, C. Cooperation of demand response and traditional power generations for providing spinning reserve. *Energy Procedia*, **2017**, *142*, 2035–2041.

Zhanga, Y., Hana, X., Yanga, M., Xub, B., Zhaoa, Y., & Zhaic, H. Adaptive robust unit commitment considering distributional uncertainty. *Electrical Power and Energy Systems*, **2019**, *104*, 635–644.

CHAPTER 2

Power Systems Control

2.1 GENERAL

In the daily operation of power systems, the power supply and demand have to be balanced to keep the frequency deviation in a small range to avoid damages to electrical facilities, and to balance the power supply and demand with optimized control (Shangfeng et al., 2018; Xi et al., 2017a, b, 2018).

2.2 LOAD CURVES

In the electric power system, a load curve is a chart illustrating the variation of the demand/electrical load over a specific period. Generation companies/authorities use this information to plan how much power they need in the present and future to generate at any certain of time. There is no difference between the load duration curve and a load curve. The load duration curve reflects the activity of a population with respect to the electrical power consumption over a given period (Többena and Schröder, 2018; Macedo et al., 2015; Khemakhem et al., 2019).

The load curves defined as the relationship between the power loads (in MW) versus the time (Figure 2.1). It can be used in economic dispatching study, system planning, and reliability evaluation. At the same time, the load in the electric power system in any period is a stochastic process. Different models of load are established in a way starting from primary load data and based on the need to find the reliability.

It is well known that there is a number of load curves types. These types can be summarized as:

- **Hourly Load Curve:** It can be plotted for a particular day. For this purpose, the average load for a different time for the whole hour is calculated. The obtained values are plotted against time. Therefore, the hourly variations loads are important especially for the end-use customer.

- **Daily Load Curve:** It can be plotted for a particular day. For this purpose, the average load for a different time for the whole day is calculated. The obtained values are plotted against time.
- **Weekly Load Curve:** The weekly load curve is a graphical plot showing the variation for the load demand by the customers with respect to time and plotted for a week. Therefore, when the load curve is plotted for a week, it is called weekly load curve.
- **Monthly Load Curve:** It can be plotted using the daily load curve defined earlier for a particular month. For this purpose, the average load for a different time for the whole month is calculated. The obtained values are plotted against time to obtain the Monthly Load Curve. Monthly Load Curve is used to fix the rate of energy.
- **Annual (Yearly) Load Curve:** It is obtained by considering the monthly load curves, where the monthly load curve was explained earlier in the present section. The monthly load curve is of that particular year. In general, the yearly load curve is used to calculate the annual load factor. According to Figure 2.2, it is clear that the available system is on the safe side. Just in case, if the maximum load (peak value) reached the full capacity, the system will be under risk condition.

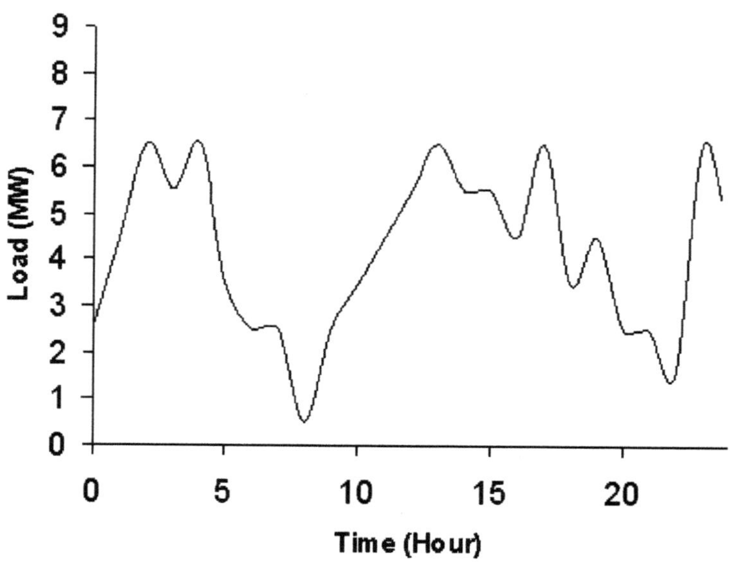

FIGURE 2.1 Hourly load curve.

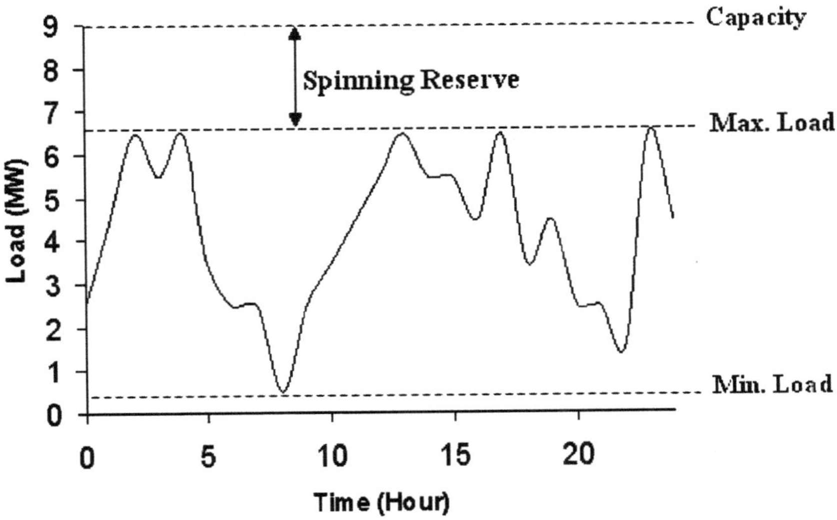

FIGURE 2.2 Typical load curve.

2.3 OBJECTIVES AND BENEFITS OF LOAD CURVES

The study of load curves reflects a number of objectives and benefits of load curves. In the present section, it is concentrating on the main points (Dubbeldam, Lin, and Schuppen, 2017). These points are:

1. The minimum load easily obtained.
2. The maximum load easily obtained.
3. The capacity easily known.
4. The spinning reserve easily calculated.
5. It will help for future planning. This means that it helps in the selection of the rating and the number of required generating units.
6. The average load easily calculated.
7. The area under the load curve (Figure 2.2) represents the energy demanded by the system (the consumed energy). From the daily load curve, we can have an insight of load at a different time for a day.

The area under the daily load curve gives the total number of units of electric energy generated. The area under this curve gives the number of units generated in a day, a week, a month, and a year (Van Dijk et al., 2017; Ma et al., 2017).

KEYWORDS

- **annual (yearly) load curve**
- **daily load curve**
- **hourly load curve**
- **load curves**
- **monthly load curve**
- **weekly load curve**

REFERENCES

Khemakhem, S., Rekik, M., & Krichen, L. Double layer home energy supervision strategies based on demand response and plug-in electric vehicle control for flattening power load curves in a smart grid. *Energy*, **2019**, *167*, 312–324.

Ma, H., Yang, Z., You, P., & Fei, M. Multi-objective biogeography-based optimization for dynamic economic emission load dispatch considering plug-in electric vehicles charging, *Energy*, **2017**, *135*, 101–111.

Macedo, M. N. Q., Galo, J. J. M. G., Almeida, L. A. L., & Lima, A. C. C. Typification of load curves for DSM in Brazil for a smart grid environment, *Electrical Power and Energy Systems*, **2015**, *67*, 216–221.

Shangfeng, X., Dajun, J., & Jinbo, W. An integrated wind power control system designing. *Energy Procedia*, **2018**, *145*, 157–162.

Többena, J., & Schröder, T. A maximum entropy approach to the estimation of spatially and sectorally disaggregated electricity load curves. *Applied Energy*, **2018**, *225*, 797–813.

Van Dijk, M. T., Wingerden, Jan, W., Ashuri, T., & Li, Y. Wind farm multi-objective wake redirection for optimizing power production and loads. *Energy*, **2017**, *121*, 561–569.

Xi, K., Dubbeldam, J. L. A., Lin, H., & Schuppen, J. H. Power-imbalance allocation control for secondary frequency control of power systems. *IFAC Papers Online*, **2017a**, *50*(1), 4382–4387.

Xi, K., Dubbeldam, J. L. A., Lin, H., & Schuppen, J. Power-imbalance allocation control for secondary frequency control of power systems. *20th World Congress IFAC Conference*. Toulouse, France, **2017b**, 4466–4471.

Xi, K., Dubbeldam, J. L. A., Lin, H. X., & Schuppen, J. H. Power-imbalance allocation control of power systems-secondary frequency control. *Automatica*, **2018**, *92*, 72–85.

CHAPTER 3

Economic Dispatch

3.1 GENERAL

Any electric power system will have several power plants. At the same time, each power plant has several generating units. At any instant of time, the total electric load in the power system is met by the generating units in different power plants. Economic dispatch control determines the power output of each generating unit within a power plant and the power plant output. The purpose of the economic dispatch study is to minimize the overall cost of fuel needed to serve the system load (McLarty et al., 2019; Zakian and Kaveh, 2018).

The economic dispatch study can be defined as the operation of generation facilities to produce energy at the lowest cost to reliably serve consumers, recognizing any operational limits of generation and transmission facilities.

It is well known that the power plants consisting of several generating units are constructed investing a huge amount of money. Fuel cost, staff salary, interest, and depreciation charges, and maintenance costs are some of the components of operating costs. The major portion of the operating cost is fuel cost, and it can be controlled. Therefore, the fuel cost shall be considered alone for further consideration.

Two constraints on Unit Commitment are needed to be considered out of the number of constraints. These two are the minimum uptime, which defined when a thermal unit is brought in; it cannot be turned off immediately. Once it is committed, it has to be in the system for a specified minimum uptime. The second constraint is the minimum downtime, which is defined when a thermal unit is de-committed; it cannot be turned on immediately. It has to remain de-committed for a specified minimum downtime.

3.2 NO LOSSES SYSTEM CONSIDERATION

The economic dispatch study (Odetayo et al., 2018) depends on:

1. Operating cost, C_i (BD/hr, \$/hr).
2. Fuel input H_i (BTU/hr), where BTU is a British thermal unit.
3. Fuel cost, f_i (BD/BTU, \$/BTU)

$$C_i = H_i \cdot f_i \text{ BD/hr}$$

4. Incremental fuel cost (Lagrange multiplier)

$$\lambda_i = \partial C_i / \partial P_i$$

where: $i = i^{th}$ generating unit; P = power (MW); λ = BD/hr. 1/MW = BD/MW.hr. The 1 US\$ = 0.378 BD.

This formula will be considered for a system that has no losses (Figure 3.1).

$$H_i = A\, P_i^3 + B\, P_i^2 + D\, P_i + E$$

Here, A, B, D, and E are constants.

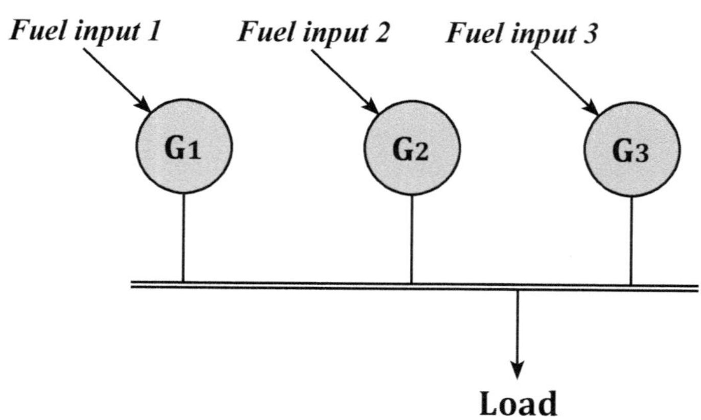

FIGURE 3.1 Three generators system.

Following the curve fitting technique (Figure 3.2), we can get the following equation:

$$\mathbf{C = H.\ f}$$

where: H = fuel input (BTU/hr); f = fuel cost (BD/BTU); and C = operating cost (BD/hr).

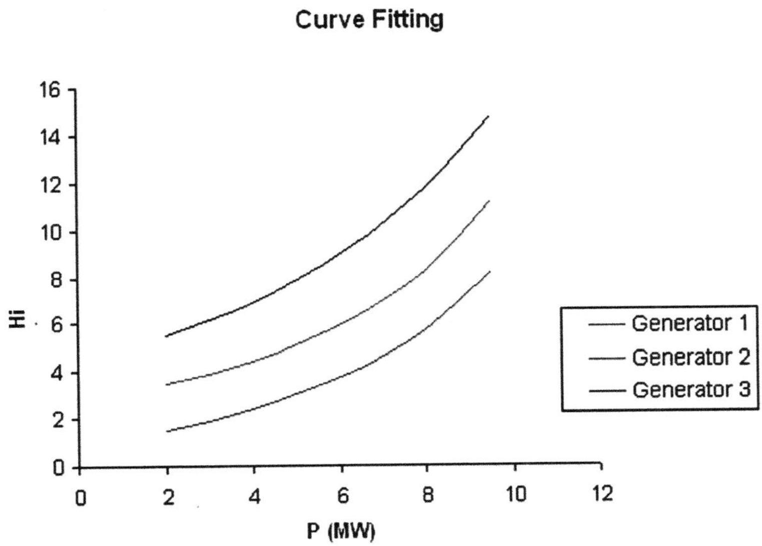

FIGURE 3.2 **(See color insert.)** Curve fitting for the three generators.

Lagrange Multiplier (λ) = Incremental Fuel Cost (BD/MW hr)

When the losses are ignored (Figure 3.3):

$$\lambda_1 = \lambda_2 = \lambda_3 = \ldots\ldots = \lambda_n$$

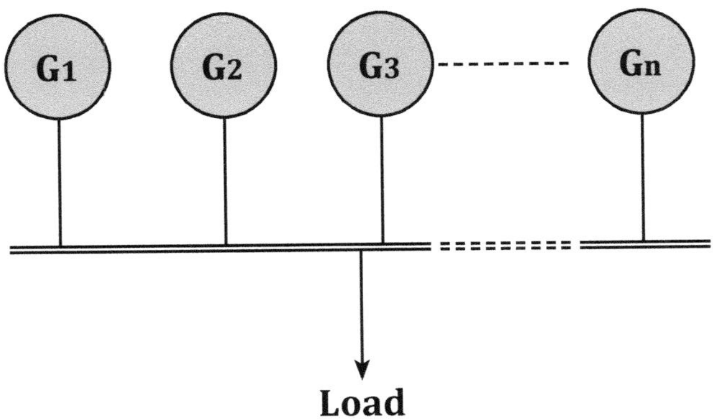

FIGURE 3.3 Generation and load system (losses ignored).

To prove the above equation, assuming that two generators considered under operation, where the losses are zero (Figure 3.4).

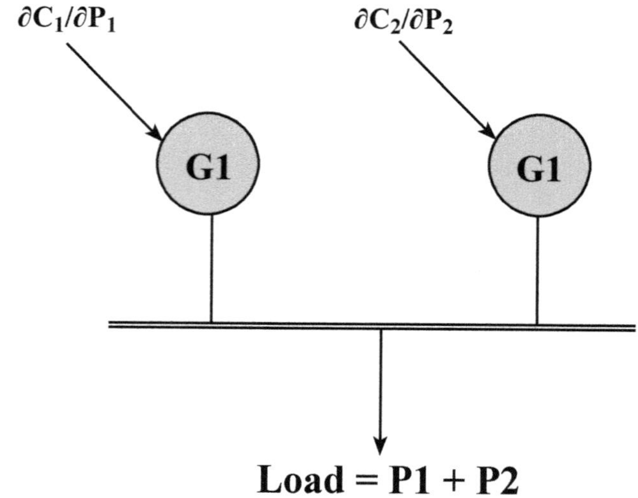

FIGURE 3.4 Two generators system with no-losses.

The total operating cost:

$$C_{Total} = C_1 + C_2 \tag{1}$$

The total generated power:

$$P_{Total} = P_1 + P_2 \tag{2}$$

Since no losses considered, therefore:

$$C_{Total} = \text{Constant and } P_{Total} = \text{Constant}$$

To find the partial differentiation of both equations (1) and (2) with respect to P_2:

For equation (1)

$$\partial CT/\partial P2 = \partial C1/\partial P2 + \partial C2/\partial P2$$

$$0 = \partial C1/\partial P1.\ \partial P1/\partial P2 + \partial C2/\partial P2 \tag{3}$$

For equation (2)

$$\partial PT/\partial P2 = \partial P1/\partial P2 + \partial\ P2/\partial P2$$

$$0 = \partial P1/\partial P2 + 1$$

$$\partial P1/\partial P2 = -1 \tag{4}$$

Substituting Eq. (4) in Eq. (3)

$$0 = \partial C_1/\partial P_1\ (-1) + \partial C_2/\partial P_2$$

$$\partial C_1/\partial P_1 = \partial C_2/\partial P_2$$

$$\lambda_1 = \lambda_2$$

- This equation applied to two generators.

$$\lambda_1 = \lambda_2 = \lambda_3 = \ldots = \lambda_n$$

- This equation applied to *n*-number of generators.

Example (3.1)

Consider the following fuel inputs for the generating units:

$$H_1 = 510 + 7.2\ P_1 + 0.00142\ P_1^2\ \text{MBTU/hr}$$

$$H_2 = 310 + 7.85\ P_2 + 0.00194\ P_2^2\ \text{MBTU/hr}$$

where the limits of power for both units are:

$$150\ \text{MW} \leq P_1 \leq 600\ \text{MW}$$

$$100\ \text{MW} \leq P_2 \leq 400\ \text{MW}$$

The fuel costs are, respectively, as follow:

$$f_1 = 1.1\ \text{BD/MBTU}$$

$$f_2 = 1.0\ \text{BD/MBTU}$$

Find the most economical solution for both units to satisfy a load of 550 MW, where the losses are ignored. Then, determine the following:

a. Fuel input.
b. Operating cost.

Solution

Since there are two units, the most economical solution has to be satisfied for the following condition:

$\lambda_1 = \lambda_2$

$C_1 = H_1 . f_1$

$= [510 + 7.2\ P_1 + 0.00142\ P_1^2][1.1]$

$= 561 + 7.92\ P_1 + 0.001562\ P_1^2$ BD/hr

$C_2 = H_2 . f_2$

$= [310 + 7.85\ P_2 + 0.001942\ P_2^2][1.0]$

$= 310 + 7.85\ P_2 + 0.00194\ P_2^2$ BD/hr

$\partial C_1/\partial P_1 = \partial C_2/\partial P_2$

$\partial C_1/\partial P_1 = 7.92 + 0.003124\ P_1$ BD/MW.hr

$\partial C_2/\partial P_2 = 7.85 + 0.00388\ P_2$ BD/MW.hr

Since $\partial C_1/\partial P_1 = \partial C_2/\partial P_2$

$7.92 + 0.003124\ P_1 = 7.85 + 0.00388\ P_2$

$0.003124\ P_1 - 0.00388\ P_2 + 7.92 - 7.85 = 0$

$0.003124\ P_1 - 0.00388\ P_2 + 0.07 = 0$

$$P1 - 1.242\ P2 + 22.407 = 0 \quad (5)$$

$P_1 + P_2 = 550$

$$P2 = 550 - P1 \quad (6)$$

Substituting (6) in (5)

$P_1 - 1.242\ (550 - P_1) + 22.407 = 0$

$P_1 - 683.1 + 1.242\ P_1 + 22.407 = 0$

$2.242\ P_1 = 683.1 - 22.407$

$P_1 = (683.1 - 22.407)/2.242$

$P_1 = 294.689$ MW

Substituting P_1 in (6)

$P_2 = 550 - 294.689$

$P_2 = 255.311$ MW

We have to check the obtained results of P_1 and P_2 if it is within limits.

In this example, it is clear that P_1 and P_2 are within the limits.

$$H_1 = 510 + 7.2\,(294.689) + 0.00142\,(294.689)^2$$
$$= 2755.0768 \text{ MBTU/hr}$$

$$H_2 = 310 + 7.85\,(255.311) + 0.00194\,(255.311)^2$$
$$= 2440.648 \text{ MBTU/hr}$$

It is clear that G_1 will consumed more fuel comparing with G_2

$$C_1 = 561 + 7.92\,(294.689) + 0.001562\,(294.689)^2$$
$$= 3030.5842 \text{ BD/hr}$$

$$C_2 = H_2 .\, f_2$$
$$= 2440.648 \text{ BD/hr}$$

$$C_{Total} = C_1 + C_2$$
$$= 5471.232 \text{ BD/hr}$$

If the value of H_1 is calculated assuming that the first generator will satisfy the needed load (which is equal to 550MW). Then,

$$H_1 = 510 + 7.2\,P_1 + 0.00142\,P_1^2 \text{ MBTU/hr}$$
$$= 510 + 7.2\,(550) + 0.00142\,(550)^2$$
$$= 4899.55 \text{ MBTU/hr}$$

$$C_1 = H_1\, f_1$$
$$= (4899.55)\,(1.1)$$
$$= 5389.505 \text{ BD/hr}$$

Comparing the results of C_{Total} "both generators are running, the cost is 5471.232 BD/hr," where the cost when running the first generator only it becomes 5389.505 BD/hr. The difference between the results is 81.73 BD/hr. Therefore, it is clear that running the first generator only it will cost less than running both.

Example (3.2)

Consider the following generating units:

$$H_1 = 510 + 7.2\,P_1 + 0.00142\,P_1^2 \text{ MBTU/hr}$$
$$H_2 = 310 + 7.85\,P_2 + 0.00194\,P_2^2 \text{ MBTU/hr}$$

where the limits of power for both units are:

$$150\ \text{MW} \le P_1 \le 600\ \text{MW}$$
$$100\ \text{MW} \le P_2 \le 400\ \text{MW}$$

The fuel costs are respectively as following:

$$f_1 = 1.1\ \text{BD/MBTU}$$
$$f_2 = 1.0\ \text{BD/MBTU}$$

Find the most economical solution for both units to satisfy a load of 600 MW, where the losses are ignored. Then, determine the following:

a. Fuel input.
b. Operating cost.

Solution

Since there are two units, the most economical solution has to be satisfied for the following condition:

$$\lambda_1 = \lambda_2$$

$$C_1 = H_1 . f_1$$

$$= [510 + 7.2\ P_1 + 0.00142\ P_1^2][1.1]$$
$$= 561 + 7.92\ P_1 + 0.001562\ P_1^2\ \text{BD/hr}$$

$$C_2 = H_2 . f_2$$

$$= [310 + 7.85\ P_2 + 0.001942\ P_2^2][1.0]$$
$$= 310 + 7.85\ P_2 + 0.00194\ P_2^2\ \text{BD/hr}$$

$$\partial C_1/\partial P_1 = \partial C_2/\partial P_2$$

$$\partial C_1/\partial P_1 = 7.92 + 0.003124\ P_1\ \text{BD/MW.hr}$$
$$\partial C_2/\partial P_2 = 7.85 + 0.00388\ P_2\ \text{BD/MW.hr}$$

Since $\partial C_1/\partial P_1 = \partial C_2/\partial P_2$

$$7.92 + 0.003124\ P_1 = 7.85 + 0.00388\ P_2$$
$$0.003124\ P_1 - 0.00388\ P_2 + 7.92 - 7.85 = 0$$
$$0.003124\ P_1 - 0.00388\ P_2 + 0.07 = 0$$

$$P1 - 1.242\ P2 + 22.407 = 0 \tag{7}$$

$P_1 + P_2 = 600$

$$P2 = 600 - P1 \quad (8)$$

Substituting (8) in (7)

$$P_1 - 1.242\,(600 - P_1) + 22.407 = 0$$
$$P_1 - 745.2 + 1.242\,P_1 + 22.407 = 0$$
$$2.242\,P_1 = 745.2 - 22.407$$
$$P_1 = (745.2 - 22.407)/2.242$$
$$P_1 = 322.388 \text{ MW}$$

Substituting P_1 in (8)

$$P_2 = 600 - 322.388$$
$$P_2 = 277.612 \text{ MW}$$

It is needed to check the obtained results of P_1 and P_2 if it is within limits or not.

In this example, it is clear that P_1 and P_2 are within limits.

$$H_1 = 510 + 7.2\,(322.388) + 0.00142\,(322.388)^2$$
$$= 2978.78 \text{ MBTU/hr}$$

$$H_2 = 310 + 7.85\,(277.612) + 0.00194\,(277.612)^2$$
$$= 2638.77 \text{ MBTU/hr}$$

It is clear that G_1 will consumed more fuel comparing with G_2

$$C_1 = 561 + 7.92\,(322.388) + 0.001562\,(322.388)^2$$
$$= 3276.66 \text{ BD/hr}$$

$$C_2 = H_2 .\, f_2$$
$$= 2638.77 \text{ BD/hr}$$

$$C_{Total} = C_1 + C_2$$
$$= 5915.43 \text{ BD/hr}$$

If the value of H_1 is calculated assuming that the first generator will satisfy the needed load (which is equal to 600 MW). Then,

$$H_1 = 510 + 7.2\,P_1 + 0.00142\,P_1^2 \text{ MBTU/hr}$$
$$= 510 + 7.2\,(600) + 0.00142\,(600)^2$$
$$= 5341.2 \text{ MBTU/hr}$$

$$C_1 = H_1 f_1$$
$$= (5341.2)(1.1)$$
$$= 5875.32 \text{ BD/hr}$$

Comparing the results of C_{Total} "both generators are running, the cost is 5915.43 BD/hr," where the cost when running the first generator only it becomes 5875.32 BD/hr. The difference between the results is 40.11 BD/hr. Therefore, it is clear that running the first generator only it will cost less than running both.

Example (3.3):

Consider the following generating units:

$$H_1 = 510 + 7.2 P_1 + 0.00142 P_1^2 \text{ MBTU/hr}$$
$$H_2 = 310 + 7.85 P_2 + 0.00194 P_2^2 \text{ MBTU/hr}$$

Where the limits of power for both units are:

$$150 \text{ MW} \leq P_1 \leq 600 \text{ MW}$$
$$100 \text{ MW} \leq P_2 \leq 400 \text{ MW}$$

The fuel costs are respectively as following:

$$f_1 = 1.0 \text{ BD/MBTU}$$
$$f_2 = 1.0 \text{ BD/MBTU}$$

Find the most economical solution for both units to satisfy a load of 600 MW, where the losses are ignored. Then, determine the following:

a. fuel input, and
b. operating cost.

Solution:

Since there are two units, the most economical solution has to be satisfy for the following condition

$$\lambda_1 = \lambda_2$$
$$C_1 = H_1 . f_1$$
$$= [510 + 7.2 P_1 + 0.00142 P_1^2][1.0]$$
$$= 510 + 7.2 P_1 + 0.00142 P_1^2 \text{ BD/hr}$$

$C_2 = H_2 . f_2$

$= [310 + 7.85\ P_2 + 0.001942\ P_2^2][1.0]$

$= 310 + 7.85\ P_2 + 0.00194\ P_2^2$ BD/hr

$\partial C_1/\partial P_1 = \partial C_2/\partial P_2$

$\partial C_1/\partial P_1 = 7.2 + 0.00284\ P_1$ BD/MW.hr

$\partial C_2/\partial P_2 = 7.85 + 0.00388\ P_2$ BD/MW.hr

Since $\partial C_1/\partial P_1 = \partial C_2/\partial P_2$

$7.2 + 0.00284\ P_1 = 7.85 + 0.00388\ P_2$

$0.00284\ P_1 - 0.00388\ P_2 + 7.2 - 7.85 = 0$

$0.00284\ P_1 - 0.00388\ P_2 + 0.65 = 0$

$$P1 - 1.366\ P2 - 228.9 = 0 \tag{9}$$

$P_1 + P_2 = 600$

$$P_2 = 600 - P_1 \tag{10}$$

Substituting (10) in (9)

$P_1 - 1.366\ (600 - P_1) - 228.9 = 0$

$P_1 - 819.6 + 1.366\ P_1 - 228.9 = 0$

$2.366\ P_1 = 819.6 - 228.9$

$P_1 = (819.6 + 228.9)/2.366$

$P_1 = 443.15$ MW

Substituting P_1 in (10)

$P_2 = 600 - 443.15$

$P_2 = 156.85$ MW

It is needed to check the obtained results of P_1 and P_2 if it is within limits or not.

In this example, it is clear that P_1 and P_2 are within limits.

$H_1 = 510 + 7.2\ (443.15) + 0.00142\ (443.15)^2$

$= 3979.5$ MBTU/hr

$H_2 = 310 + 7.85\ (156.85) + 0.00194\ (156.85)^2$

$= 1589$ MBTU/hr

It is clear that G_1 will consumed more fuel comparing with G_2

$$C_1 = 561 + 7.92\,(443.15) + 0.001562\,(443.15)^2$$
$$= 3979.5 \text{ BD/hr}$$

$$C_2 = H_2 .\, f_2$$
$$= 1589 \text{ BD/hr}$$

$$C_{Total} = C_1 + C_2$$
$$= 5568.5 \text{ BD/hr}$$

If the value of H_1 is calculated assuming that the first generator will satisfy the needed load (which is equal to 600MW). Then,

$$H_1 = 510 + 7.2\,P_1 + 0.00142\,P_1^2 \text{ MBTU/hr}$$
$$= 510 + 7.2\,(600) + 0.00142\,(600)^2$$
$$= 5341.2 \text{ MBTU/hr}$$

$$C_1 = H_1\, f_1$$
$$= (5341.2)\,(1.0)$$
$$= 5341.2 \text{ BD/hr}$$

Comparing the results of C_{Total} "both generators are running, the cost is 5568.5 BD/hr," where the cost when running the first generator only it becomes 5341.2 BD/hr. The difference between the results is 227.3 BD/hr. Therefore, it is clear that running the first generator only it will cost less than running both.

Example (3.4):

Consider the following generating units:

$$H_1 = 510 + 7.2\,P_1 + 0.00142\,P_1^2 \text{ MBTU/hr}$$
$$H_2 = 310 + 7.85\,P_2 + 0.00194\,P_2^2 \text{ MBTU/hr}$$
$$H_3 = 78 + 7.97\,P_3 + 0.00482\,P_3^2 \text{ MBTU/hr}$$

Where the limits of power for both units are:

$$150 \text{ MW} \leq P_1 \leq 600 \text{ MW}$$
$$100 \text{ MW} \leq P_2 \leq 400 \text{ MW}$$
$$50 \text{ MW} \leq P_3 \leq 200 \text{ MW}$$

The fuel costs are respectively as following:

$f_1 = 1.1$ BD/MBTU
$f_2 = 1.0$ BD/MBTU
$f_3 = 1.2$ BD/MBTU

Find the most economical solution for the three units to satisfy a load of 900 MW, where the losses are ignored. Then, determine C_1, C_2, and C_3.

Solution:
Since there are two units, the most economical solution has to be satisfied for the following condition:

$$\lambda_1 = \lambda_2 = \lambda_3$$

$$C_1 = H_1 . f_1$$
$$= [510 + 7.2\ P_1 + 0.00142\ P_1^2][1.1]$$
$$= 561 + 7.92\ P_1 + 0.001562\ P_1^2 \text{ BD/hr}$$

$$C_2 = H_2 . f_2$$
$$= [310 + 7.85\ P_2 + 0.001942\ P_2^2][1.0]$$
$$= 310 + 7.85\ P_2 + 0.00194\ P_2^2 \text{ BD/hr}$$

$$C_3 = H_3 . f_3$$
$$= [78 + 7.97\ P_3 + 0.00482\ P_3^2][1.2]$$
$$= 93.6 + 9.564\ P_3 + 0.005784\ P_3^2 \text{ BD/hr}$$

$$\partial C_1/\partial P_1 = \partial C_2/\partial P_2 = \partial C_3/\partial P_3$$

$$\partial C_1/\partial P_1 = 7.92 + 0.003124\ P_1 \text{ BD/MWhr}$$
$$\partial C_2/\partial P_2 = 7.85 + 0.00388\ P_2 \text{ BD/MWhr}$$
$$\partial C_3/\partial P_3 = 9.564 + 0.01157\ P_3 \text{ BD/MWhr}$$

For the most economical solution: $\partial C_1/\partial P_1 = \partial C_2/\partial P_2 = \partial C_3/\partial P_3$

Since $\partial C_1/\partial P_1 = \partial C_2/\partial P_2$

$$7.92 + 0.003124\ P_1 = 7.85 + 0.00388\ P_2$$
$$0.003124\ P_1 - 0.00388\ P_2 + 7.92 - 7.85 = 0$$
$$0.003124\ P_1 - 0.00388\ P_2 + 0.07 = 0$$
$$P_1 - 1.242\ P_2 + 22.41 = 0 \quad (11)$$

$\partial C_1/\partial P_1 = \partial C_3/\partial P_3$

$$7.92 + 0.003124\ P_1 = 9.564 + 0.01157\ P_3$$
$$0.003124\ P_1 - 0.01157\ P_3 + 7.92 - 9.564 = 0$$
$$0.003124\ P_1 - 0.01157\ P_3 - 1.644 = 0$$

$$P_1 - 3.7\ P_3 - 526.25 = 0 \tag{12}$$

$$P_1 + P_2 + P_3 = 900$$

$$P_3 = 900 - P_1 - P_2 \tag{13}$$

To solve the three equations (11), (12), and (13): the Gauss Elimination method is applied to find the values of the output powers. The following values are obtained:

$$P_1 = 493.48\ \text{MW}$$
$$P_2 = 415.37\ \text{MW}$$
$$P_3 = -\ 8.855\ \text{MW}$$

Since two values obtained out of the limits (P_2 and P_3), where the third has a negative sign. Therefore, the answers should be adjusted based on the following limits:

$$150\ \text{MW} \leq P_1 \leq 600\ \text{MW}$$
$$100\ \text{MW} \leq P_2 \leq 400\ \text{MW}$$
$$50\ \text{MW} \leq P_3 \leq 200\ \text{MW}$$

The answers of P_3 can be adjusted by adding +60MW. Therefore, the final solution will be:

$$P_1 = 493.48\ \text{MW}$$
$$P_2 = 355.37\ \text{MW}$$
$$P_3 = 51.145\ \text{MW}$$

And the values of C's are:

$$C_1 = 4849.74\ \text{BD/hr}$$
$$C_2 = 3344.65\ \text{BD/hr}$$
$$C_3 = 597.88\ \text{BD/hr}$$

Example (3.5):

Consider the following generating units:

$$H_1 = 510 + 7.20\,P_1 + 0.00142\,P_1^2 \text{ MBTU/hr}$$
$$H_2 = 310 + 7.85\,P_2 + 0.00194\,P_2^2 \text{ MBTU/hr}$$
$$H_3 = 78 + 7.97\,P_3 + 0.00482\,P_3^2 \text{ MBTU/hr}$$

Where the limits of power for both units are:

$$150 \text{ MW} \leq P_1 \leq 600 \text{ MW}$$
$$100 \text{ MW} \leq P_2 \leq 400 \text{ MW}$$
$$50 \text{ MW} \leq P_3 \leq 200 \text{ MW}$$

The fuel costs are respectively as following:

$$f_1 = 1.1 \text{ BD/MBTU}$$
$$f_2 = 1.0 \text{ BD/MBTU}$$
$$f_3 = 1.2 \text{ BD/MBTU}$$

Find the most economical solution for the three units to satisfy a load of 1050 MW. The losses ignored. Then, determine C_1, C_2, and C_3.

Solution:

Since there are two units, the most economical solution has to be satisfied for the following condition:

$$\lambda_1 = \lambda_2 = \lambda_3$$

$$C_1 = H_1 . f_1$$
$$= [510 + 7.2\,P_1 + 0.00142\,P_1^2][1.1]$$
$$= 561 + 7.92\,P_1 + 0.001562\,P_1^2 \text{ BD/hr}$$

$$C_2 = H_2 . f_2$$
$$= [310 + 7.85\,P_2 + 0.001942\,P_2^2][1.0]$$
$$= 310 + 7.85\,P_2 + 0.00194\,P_2^2 \text{ BD/hr}$$

$$C_3 = H_3 . f_3$$
$$= [78 + 7.97\,P_3 + 0.00482\,P_3^2][1.2]$$
$$= 93.6 + 9.564\,P_3 + 0.005784\,P_3^2 \text{ BD/hr}$$

$\partial C_1/\partial P_1 = \partial C_2/\partial P_2 = \partial C_3/\partial P_3$

$$\partial C_1/\partial P_1 = 7.92 + 0.003124\ P_1\ \text{BD/MWhr}$$
$$\partial C_2/\partial P_2 = 7.85 + 0.00388\ P_2\ \text{BD/MWhr}$$
$$\partial C_3/\partial P_3 = 9.564 + 0.01157\ P_3\ \text{BD/MWhr}$$

For the most economical solution: $\partial C_1/\partial P_1 = \partial C_2/\partial P_2 = \partial C_3/\partial P_3$

Since $\partial C_1/\partial P_1 = \partial C_2/\partial P_2$

$$7.92 + 0.003124\ P_1 = 7.85 + 0.00388\ P_2$$
$$0.003124\ P_1 - 0.00388\ P_2 + 7.92 - 7.85 = 0$$
$$0.003124\ P_1 - 0.00388\ P_2 + 0.07 = 0$$
$$P_1 - 1.242\ P_2 + 22.41 = 0 \tag{14}$$

$\partial C_1/\partial P_1 = \partial C_3/\partial P_3$

$$7.92 + 0.003124\ P_1 = 9.564 + 0.01157\ P_3$$
$$0.003124\ P_1 - 0.01157\ P_3 + 7.92 - 9.564 = 0$$
$$0.003124\ P_1 - 0.01157\ P_3 - 1.644 = 0$$
$$P_1 - 3.7\ P_3 - 526.25 = 0 \tag{15}$$
$$P_1 + P_2 + P_3 = 1050$$
$$P_3 = 1050 - P_1 - P_2 \tag{16}$$

To solve the three equations (14), (15), and (16): the Gauss Elimination method is applied to find the values of the output powers. The following values obtained:

$$P_1 = 565.76\ \text{MW}$$
$$P_2 = 473.56\ \text{MW}$$
$$P_3 = 10.68\ \text{MW}$$

Since the second generator exceeded the limit by 73.56 MW and the third has less than the minimum power output value. Therefore, 73.56MW can be added to the third generator, which is going to be 84.24MW. In this case, the powers will be in the range by adjustment.

$$150\ \text{MW} \le P_1 \le 600\ \text{MW}$$
$$100\ \text{MW} \le P_2 \le 400\ \text{MW}$$
$$50\ \text{MW} \le P_3 \le 200\ \text{MW}$$

The final solution will be:

$$P_1 = 565.76 \text{ MW}$$
$$P_2 = 400 \text{ MW}$$
$$P_3 = 84.24 \text{ MW}$$

And the values of C's are:

$$C_1 = 5541.79 \text{ BD/hr}$$
$$C_2 = 3760.4 \text{ BD/hr}$$
$$C_3 = 940.317 \text{ BD/hr}$$

3.3 LOSSES SYSTEM CONSIDERATION

If we have a system having a number of generators supplying a load (Jordehi, 2018) through a transmission line (Figure 3.5):

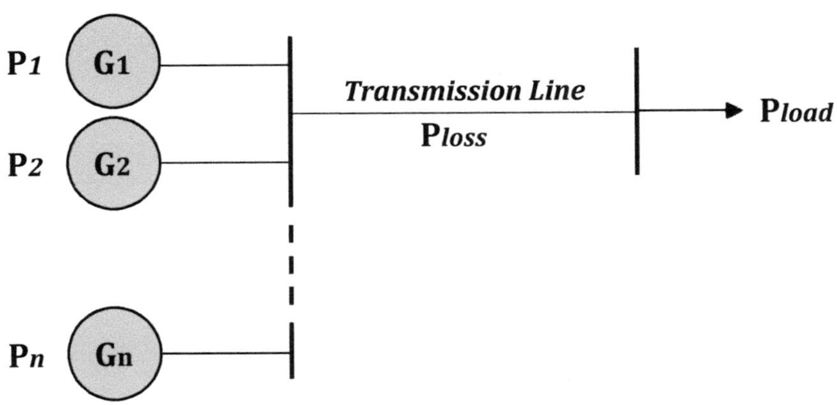

FIGURE 3.5 Power systems with losses considered.

$$\sum_{i=1}^{n} P_i = P_{load} + P_{loss}$$

$$\Phi = P_{load} + P_{loss} - \sum_{i=1}^{n} P_i = 0$$

$L = CT + \lambda\,\Phi$

$$L = CT + \lambda\left[P_{load} + P_{loss} - \sum_{i=1}^{n} P_i\right]$$

For i^{th} generator:

$$L = C_i + \lambda\,[P_{load} + P_{loss} - P_i]$$
$$L = C_i + \lambda\,P_{load} + \lambda\,P_{loss} - \lambda\,P_i$$

For the most economical solution:

$$\partial L/\partial P_i = \partial C_i/\partial P_i + \lambda\,\partial P_{load}/\partial P_i + \lambda\,\partial P_{loss}/\partial P_i - \lambda\,\partial P_i/\partial P_i = 0$$

We are looking for a specific load P_{load}

$$\partial C_i/\partial P_i + \lambda\,[0] + \lambda\,\partial P_{loss}/\partial P_i - \lambda\,(1) = 0$$

$$\partial C_i/\partial P_i = \lambda\,[1 - \partial P_{loss}/\partial P_i]$$

There are some terms used for P_{Load}:

$$P_{load} = P_{demand} = P_{required}$$

As a summary for both (losses and No Losses System)

1. **No Losses System:**

$$\lambda_1 = \lambda_2 = \ldots\ldots = \lambda_n$$
$$\partial C_1/\partial P_1 = \partial C_2/\partial P_2 = \ldots = \partial C_n/\partial P_n$$

2. **Losses Included:**

$$\partial C_i/\partial P_i = \lambda\,[1 - \partial P_{loss}/\partial P_i]$$
$$P_{loss} = \alpha\,P_1^2 + \beta\,P_1 P_2 + \gamma\,P_2^2$$

where α, β, and γ are constants. This case will consider only two generators supplying a load.

P_{loss} might be written in the following form:

$$P_{loss} = \alpha\,P_1^2 + \beta\,P_2^2 + \gamma\,P_3^2 + \ldots\ldots$$

Example (3.6):

Consider the following example which has three different types of generators:

$$H_1 = 550 + 7.20\ P_1 + 0.00142\ P_1^2 \ \text{MBTU/hr}$$
$$H_2 = 310 + 7.85\ P_2 + 0.00194\ P_2^2 \ \text{MBTU/hr}$$
$$H_3 = 78 + 7.97\ P_3 + 0.00482\ P_3^2 \ \text{MBTU/hr}$$

$f_1 = 1.1$ BD/MBTU
$f_2 = 1.0$ BD/MBTU
$f_3 = 1.0$ BD/MBTU

$$P_{loss} = 0.00003\ P_1^2 + 0.00009\ P_2^2 + 0.00012\ P_3^2 \ \text{MW}$$

The required load is 850 MW, and the limits are:

$$150\text{MW} \le P_1 \le 600 \text{ MW}$$
$$100\text{MW} \le P_2 \le 400 \text{ MW}$$
$$50\text{MW} \le P_3 \le 300 \text{ MW}$$

Calculate P_1, P_2, P_3, and P_{Loss} for the above system. Then find C_1, C_2, and C_3. Their iterations are required.

Solution:

Assuming that the starting values for:

$P_1 = 400$ MW, $P_2 = 250$ MW and $P_3 = 200$ MW

$$C_1 = H_1 .\ f_1$$
$$= 605 + 7.92\ P_1 + 0.001562\ P_1^2 \quad \text{BD/hr}$$
$$C_2 = 310 + 7.850\ P_2 + 0.00194\ P_2^2 \quad \text{BD/hr}$$
$$C_3 = 78 + 7.97\ P_3 + 0.00482\ P_3^2 \quad \text{BD/hr}$$

$$\partial C_1 / \partial P_1 = 7.92 + 0.003124\ P_1 \quad \text{BD/hr}$$
$$\partial C_2 / \partial P_2 = 7.85 + 0.00388\ P_2 \quad \text{BD/hr}$$
$$\partial C_3 / \partial P_3 = 7.97 + 0.00964\ P_3 \quad \text{BD/hr}$$

$$P_{loss} = 0.00003\ P_1^2 + 0.00009\ P_2^2 + 0.00012\ P_3^2 \ \text{MW}$$

$$\partial P_{loss} / \partial P_1 = 0.00006\ P_1$$
$$\partial P_{loss} / \partial P_2 = 0.00018\ P_2$$
$$\partial P_{loss} / \partial P_3 = 0.00024\ P_3$$
$$\partial C_i / \partial P_i = \lambda\ [1 - \partial P_{loss} / \partial P_i]$$

First Iteration:

$$7.92 + 0.003124\ P_1 = \lambda\ [1 - 0.00006\ P_1]$$
$$7.85 + 0.003880\ P_2 = \lambda\ [1 - 0.00018\ P_2]$$
$$7.97 + 0.009640\ P_3 = \lambda\ [1 - 0.00024\ P_3]$$

Substituting P_1:

$$7.92 + 0.003124\ P_1 = \lambda\ [1 - 0.00006\ (400)] = 0.976\ \lambda$$

Substituting P_2:

$$7.85 + 0.003880\ P_2 = \lambda\ [1 - 0.00018\ (250)] = 0.955\ \lambda$$

Substituting P_3:

$$7.97 + 0.009640\ P_3 = \lambda\ [1 - 0.00024\ (200)] = 0.952\ \lambda$$

$$P_{loss} = 0.00003\ (400)^2 + 0.00009\ (250)^2 + 0.00012\ (200)^2 = 15.225\ \text{MW}$$

$$P_1 + P_2 + P_3 = P_{Load} + P_{Loss} = 850 + 15.225 = 865.225\ \text{MW}$$

$$P_1 = (0.976\ \lambda - 7.92)/0.003124 \quad (17)$$
$$P_2 = (0.955\ \lambda - 7.85)/0.003880 \quad (18)$$
$$P_3 = (0.952\ \lambda - 7.97)/0.009640 \quad (19)$$

Adding (17), (18), and (19), we get:

$$P_1 + P_2 + P_3 = [(0.976\ \lambda - 7.92)/0.003124] + [(0.955\ \lambda - 7.85)/0.003880] + [(0.952\ \lambda - 7.97)/0.009640]$$
$$865.225 = 657.3093\ \lambda - 5385.171$$
$$6250.396 = 657.3093\ \lambda$$

$$\lambda = 9.509065\ \text{BD/MWhr}$$

Substituting the obtained value of $\lambda = 9.509065$BD/MWhr to find P_1, P_2, and P_3.

$$P_1 = [0.976\,(9.509065) - 7.92]/0.003124 = 435.6106 \text{ MW}$$
$$P_2 = [0.955\,(9.509065) - 7.85]/0.003880 = 317.3086 \text{ MW}$$
$$P_3 = [0.952\,(9.509065) - 7.97]/0.009640 = 112.3060 \text{ MW}$$

	P_1 (MW)	P_2 (MW)	P_3 (MW)	P_{Loss} (MW)	λ (BD/MWhr)
1st Iteration	435.6106	317.3086	112.306	15.225	9.509065
2nd Iteration	515.2065	354.734	–3.6729	16.268	9.785259
3rd Iteration	435.6105	317.3085	112.30599	15.225	9.785259

Try starting values:

$$P_1 = 400 \text{ MW}$$
$$P_2 = 250 \text{ MW}$$
$$P_3 = 200 \text{ MW}$$
$$P_{load} = 850 \text{ MW}$$

Example (3.7):

Consider the following example. It has three different types of generators:

$$H_1 = 210 + 3.0\,P_1 + 0.00071\,P_1^2 \quad \text{MBTU/hr}$$
$$H_2 = 150 + 3.5\,P_2 + 0.00045\,P_2^2 \quad \text{MBTU/hr}$$
$$H_3 = 40 + 3.4\,P_3 + 0.00021\,P_3^2 \quad \text{MBTU/hr}$$

$$f_1 = 1.20 \quad \text{BD/MBTU}$$
$$f_2 = 1.10 \quad \text{BD/MBTU}$$
$$f_3 = 1.15 \quad \text{BD/MBTU}$$

$$P_{loss} = 0.00003\,P_1^2 + 0.00009\,P_2^2 + 0.00012\,P_3^2 \quad \text{MW}$$

The required load is 500 MW and the limits are:

$$150\text{MW} \le P_1 \le 300 \text{ MW}$$
$$100\text{MW} \le P_2 \le 200 \text{ MW}$$
$$50\text{MW} \le P_3 \le 150 \text{ MW}$$

Calculate P_1, P_2, P_3 and P_{loss} for the above system. One iteration is required, where the starting values are P_1 = 200 MW, P_2= 160 MW and P_3 = 140 MW. At least four decimal points are required.

Solution:

Assuming that the starting values for:

$P_1 = 200$ MW, $P_2 = 160$ MW and $P_3 = 140$ MW

$C_1 = H_1 . f_1$

$= 252 + 3.6\ P_1 + 0.000852\ P_1^2$ BD/hr

$C_2 = 165 + 3.85\ P_2 + 0.000495\ P_2^2$ BD/hr

$C_3 = 46 + 3.91\ P_3 + 0.000242\ P_3^2$ BD/hr

$\partial C_1 / \partial P_1 = 3.60 + 0.001704\ P_1$ BD/hr

$\partial C_2 / \partial P_2 = 3.85 + 0.000990\ P_2$ BD/hr

$\partial C_3 / \partial P_3 = 3.91 + 0.000483\ P_3$ BD/hr

$P_{Loss} = 0.00003\ P_1^2 + 0.00009\ P_2^2 + 0.00012\ P_3^2$ MW

$\partial P_{Loss} / \partial P_1 = 0.00006\ P_1$

$\partial P_{Loss} / \partial P_2 = 0.00018\ P_2$

$\partial P_{Loss} / \partial P_3 = 0.00024\ P_3$

$\partial C_i / \partial P_i = \lambda\ [1 - \partial P_{loss} / \partial P_i]$

First Iteration:

$3.60 + 0.001704\ P_1 = \lambda\ [1 - 0.00006\ P_1]$

$3.85 + 0.000990\ P_2 = \lambda\ [1 - 0.00018\ P_2]$

$3.91 + 0.000483\ P_3 = \lambda\ [1 - 0.00024\ P_3]$

Substituting P_1:

$3.60 + 0.001704\ P_1 = \lambda\ [1 - 0.00006\ (200)]$

$= 0.988\ \lambda$

Substituting P_2:

$3.85 + 0.000990\ P_2 = \lambda\ [1 - 0.00018\ (160)]$

$= 0.9712\ \lambda$

Substituting P_3:

$3.91 + 0.000483\ P_3 = \lambda\ [1 - 0.00024\ (140)]$

$= 0.9664\ \lambda$

$P_{Loss} = 0.00003\ (200)^2 + 0.00009\ (160)^2 + 0.00012\ (140)^2$

$= 5.856$ MW

$$P_1 + P_2 + P_3 = P_{Load} + P_{Loss}$$
$$= 500 + 5.856$$
$$= 505.856 \text{ MW}$$

$$P_1 = (0.9880\,\lambda - 3.60)/0.001704 \quad (20)$$
$$P_2 = (0.9712\,\lambda - 3.85)/0.000990 \quad (21)$$
$$P_3 = (0.9664\,\lambda - 3.91)/0.000483 \quad (22)$$

Adding (20), (21), and (22), we get:

$$P_1 + P_2 + P_3 = [(0.9880\,\lambda - 3.60)/0.001704] + [(0.9712\,\lambda - 3.85)/0.000990]$$
$$+ [(0.9664\,\lambda - 3.91)/0.000483]$$
$$505.856 = 3561.650465\,\lambda - 14096.80304$$
$$14602.65904 = 3561.650465\lambda$$

$$\lambda = 4.099969715 \text{ BD/MWhr}$$

Substituting the obtained value of $\lambda = 4.099969715$ BD/MWhr to find P_1, P_2, and P_3.

$$P_1 = [0.9880\,(4.099969715) - 3.60]/0.001704 = 264.5364308 \quad (23)$$
$$P_2 = [0.9712\,(4.099969715) - 3.85]/0.000990 = 133.222815 \quad (24)$$
$$P_3 = [0.9664\,(4.099969715) - 3.91]/0.000483 = 108.0967542 \quad (25)$$

	P_1 (MW)	P_2 (MW)	P_3 (MW)	P_{Loss} (MW)	λ (BD/MWhr)
1st Iteration	264.5364308	133.222815	108.0967542	5.856	4.0999697
2nd Iteration	316.0970096	253.634294	64.63238048	5.099	9.05114857
3rd Iteration	435.6105264	317.308485	112.3059884	15.225	9.50906484

3.4 SPINNING RESERVE AND POWER INTERCHANGE

A question might be raised what does the spinning reserve means?

The answer to the question is that the reserve power stored in the generating unit as extra power. Therefore, the spinning reserve is the extra generating capacity, which is available by increasing the output power of generators (Hreinsson et al., 2015; Palacio et al., 2015; Wang et al., 2017; Hirasea et al., 2018). These generators are already connected to the power system network (Figure 3.6).

$$\text{Spinning Reserve} = \sum_{i=1}^{n} \text{Capacity of the units} - P_{loss} - P_{load}$$

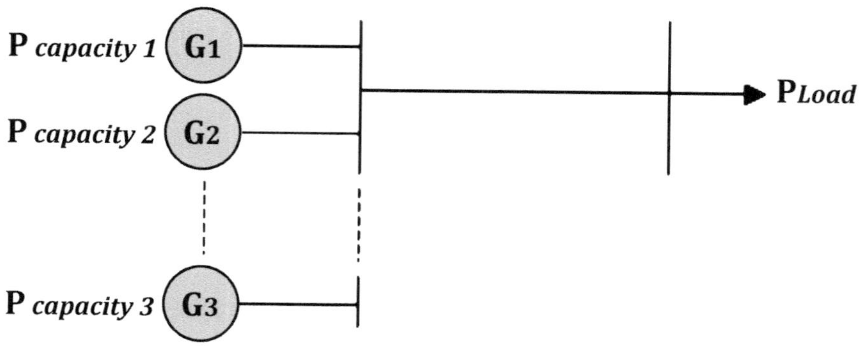

FIGURE 3.6 Power systems with the losses consideration.

$$\sum_{i=1}^{n} P_i = P_{load} + P_{loss}$$

Another question is that what does power interchange means?

The answer will be that the interchange power is the transmitted energy from one point (Region) to another point through a transmission line based on the need (Figure 3.7).

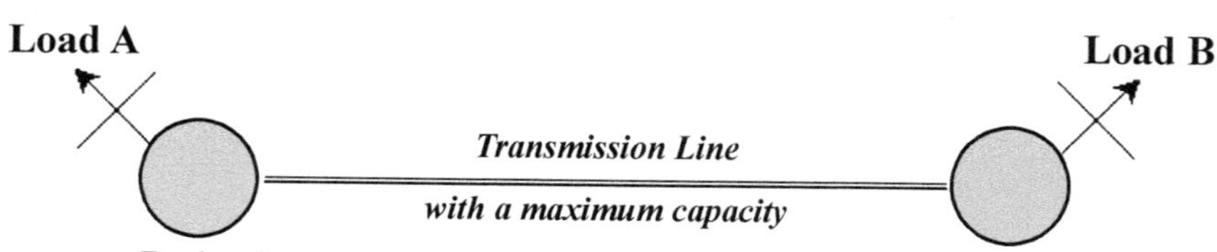

FIGURE 3.7 Tie-line system.

Example (3.8):

Consider two regions (Region I & II). The first region has three generating-units, where the second has only two units.

The maximum capacities of the first region units are:

Unit # 1 = 1000 MW

Unit # 2 = 800 MW

Unit # 3 = 800 MW

The second region maximum capacities of units are:

Unit # 4 = 1200 MW

Unit # 5 = 600 MW

a. Fill the given table (i.e., Table 3.1) and find how much interchange needs to be transferred? Where both regions connected through 550 MW maximum capacity transmission line.

TABLE 3.1 Results for Part "a"

Region	Unit	Unit Capacity	Unit Output	Region Generation	Spinning Reserve	Region Load	Interchange Power
I	1	1000MW	900 MW	1740 MW	100 MW	1900 MW	–160 MW Out of Service
	2	800 MW	420 MW		380 MW		
	3	800 MW	420 MW		380 MW		
II	4	1200 MW	1040 MW	1350 MW	160 MW	1190 MW	+160 MW In Service
	5	600 MW	310 MW		290 MW		

It can be concluded that the power of 160 MW can be transferred from Region II to Region I. As a note, we have taken in our consideration the maximum capacity of the transmission line.

b. Fill the given table (i.e., Table 3.2), if both regions are connected by 100 MW maximum capacity transmission line:

TABLE 3.2 Results for Part "b"

Region	Unit	Unit Capacity	Unit Output	Region Generation	Spinning Reserve	Region Load	Interchange Power
I	1	1000MW	900 MW	1800 MW	100 MW	1900 MW	–100 MW Out of Service
	2	800 MW	450 MW		350 MW		
	3	800 MW	450 MW		350 MW		
II	4	1200 MW	1040 MW	1350 MW	160 MW	1190 MW	+160 MW In Service
	5	600 MW	310 MW		290 MW		

Example (3.9):

Refer to the previous example "Example 8" regarding the interchange power, where the tie-line capacity is 550 MW:

a. Fill Table 3.3 and consider that unit # 4 is lost, and unit #5 has to be run at full capacity. Find the interchange power between both regions.
b. Fill Table 3.4 when unit # 4 is back to the service, but unit # 2 is lost. How can we solve the problem with the remained units in both regions?

Solution:
Case a.

TABLE 3.3 Case "a"

Region	Unit	Unit Capacity	Unit Output	Region Generation	Spinning Reserve	Region Load	Interchange Power
	1	1000MW	950 MW	2450 MW	50 MW	1900 MW	+550 MW
I	2	800 MW	750 MW		50 MW		In Service
	3	800 MW	750 MW		50 MW		
II	4	~~1200 MW~~		600 MW		1150 MW	–40 MW
	5	600 MW	600 MW		0 MW		Out of Service

Case b.

TABLE 3.4 Case "b"

Region	Unit	Unit Capacity	Unit Output	Region Generation	Spinning Reserve	Region Load	Interchange Power
	1	1000MW	950 MW	1400 MW	50 MW	1900 MW	–500 MW
I	2	~~800 MW~~					Out of Service
	3	800 MW	450 MW		350 MW		
II	4	1200 MW	1040 MW	1350 MW	160 MW	1190 MW	+160 MW
	5	600 MW	310 MW		290 MW		In Service

Table 3.5 shows the final solution for Example 3.9.

TABLE 3.5 The Final Solution for the System "Example 3.9"

Region	Unit	Unit Capacity	Unit Output	Region Generation	Spinning Reserve	Region Load	Interchange Power
I	1	1000MW	950 MW	1740 MW	50 MW	1900 MW	–160 MW Out of Service
	2	~~800 MW~~					
	3	800 MW	450 MW		10 MW		
II	4	1200 MW	1040 MW	1350 MW	160 MW	1190 MW	+160 MW In Service
	5	600 MW	310 MW		290 MW		

Example (3.10):

Assuming that we have two regions A and B. A power is needed to be transferred from one to the other; they are connected by tie-line having a capacity equal to 600 MW.

Find the interchange power for both regions. Then, which of them will transfer the power to the other? And how much power will be transferred?

Solution: (Tables 3.11 and 3.12)

TABLE 3.6 Regions (A and B) Power Systems

Region	Unit	Unit Capacity	Unit Output	Region Generation	Spinning Reserve	Region Load	Interchange Power
A	1	1000MW	~~942.308 MW~~	~~2450 MW~~	~~57.692 MW~~	1900 MW	~~+550 MW~~ In Service
	2	800 MW	753.846 MW		46.154 MW		
	3	800 MW	753.846 MW		46.154 MW		
B	4	400 MW	400 MW	600 MW	0 MW	1190 MW	–590 MW Out of Service
	5	200 MW	200 MW		0 MW		

TABLE 3.7 Regions A and B

Region	Unit	Unit Capacity	Unit Output	Region Generation	Spinning Reserve	Region Load	Interchange Power
A	1	1000MW	982.308 MW	2490 MW	17.892MW	1900 MW	+550 MW In service
	2	800 MW	753.864 MW		46.154 MW		
	3	800 MW	753.864 MW		46.154 MW		
B	4	400 MW	400 MW	600 MW	0 MW	1190 MW	–590 MW Out of service
	5	200 MW	200 MW		0 MW		

At this stage, the tie-line capacity is 600 MW, and the interchange power of the regions is:

Region A: +590 MW after increasing 1st unit output to 982.308 MW instead of 942.308 MW

Region B: –590 MW. This amount of power transferred from region A to Region B without any problem.

TABLE 3.8 Regions (A and B)

Region	Unit	Unit Capacity	Unit Output	Region Generation	Spinning Reserve	Region Load	Interchange Power
A	1	1000MW	942.308MW	2450 MW	57.692 MW	1900 MW	+550 MW In Service
	2	800 MW	753.846 MW		46.154 MW		
	3	800 MW	753.846 MW		46.154 MW		
B	4	460 MW	~~400 MW~~	~~600 MW~~	~~60 MW~~	1190 MW	~~–590 MW~~ Out of Service
	5	250 MW	~~200MW~~		~~50 MW~~		

TABLE 3.9 Regions A and B

Region	Unit	Unit Capacity	Unit Output	Region Generation	Spinning Reserve	Region Load	Interchange Power
A	1	1000MW	942.308 MW	2450 MW	57.692 MW	1900 MW	+550 MW In Service
	2	800 MW	753.846 MW		46.154 MW		
	3	800 MW	753.846 MW		46.154 MW		
B	4	460 MW	420 MW	640 MW	40 MW	1190 MW	–550 MW Out of Service
	5	250 MW	220 MW		30 MW		

In this case:

The inter charge power of region A = +550 MW

The inter charge power of region B = –550 MW

The total power of 550 MW can be transferred from region A to region B through the tie-line, where the output power of Unit 4 = 420 MW and of Unit 5 = 220 MW.

3.5 SIMPLE PEAK VALLEY CURVE

Figure 3.8 shows the simple peak valley curve (Zimmer and Gabet, 2018).

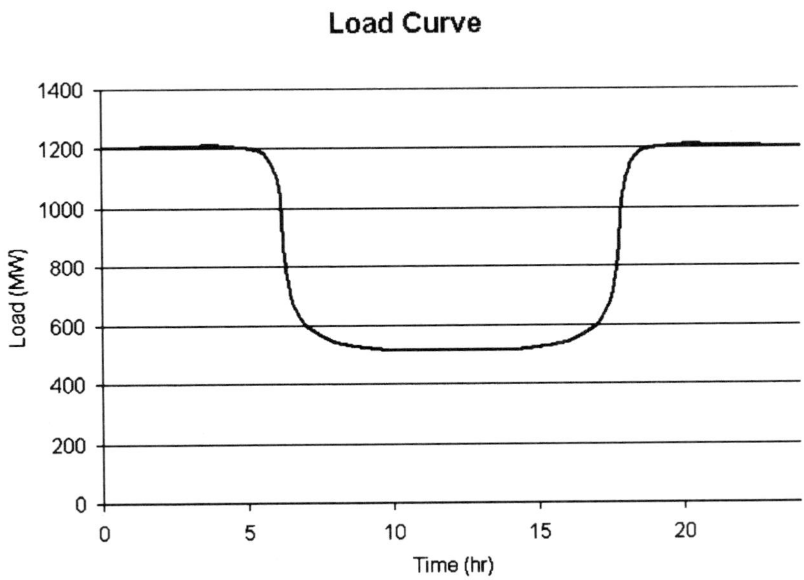

FIGURE 3.8 Simple peak valley curve.

Unit # 1	$150\ \text{MW} \leq P_1 \leq 600\ \text{MW}$
Unit # 2	$100\ \text{MW} \leq P_2 \leq 400\ \text{MW}$
Unit # 3	$50\ \text{MW} \leq P_3 \leq 200\ \text{MW}$

It is well known that the load curve is the relationship between the load and the time. Therefore, the above load-curve represents the variation of the load when we operate their different generating units for one day. This curve is known as the simple peak-valley curve.

Consider Table 3.10, where the operated units are 1, 2, and 3. If we are looking for a specific load and trying to find the most suitable economical solution for the problem:

1. We have to know the required load.
2. Finding possible solutions.
3. Following the Economic Dispatch Technique for possible solutions.

4. Calculating the total cost of the obtained solutions.
5. Choosing the most economical solution based on the obtained results.

TABLE 3.10 Operation of Three Units

Load (MW)	Unit # 1	Unit # 2	Unit # 3
1200	ON 600 MW	ON 400 MW	ON 200 MW
1100	ON 600 MW	ON 350 MW	ON 150 MW
1000	ON 600 MW	ON 400 MW	OFF
900	ON 550 MW	ON 350 MW	OFF
800	ON 600 MW	OFF	ON 200 MW
700	ON 500 MW	ON 200 MW	OFF
600	ON 600 MW	OFF	OFF
500	ON 500 MW	OFF	OFF
400	OFF	ON 400 MW	OFF
300	OFF	ON 300 MW	OFF
200	OFF	OFF	ON 200 MW
100	OFF	ON 100 MW	OFF

Example (3.11):

Three generators are under operation, where their ratings are as follows:

$$250\text{ MW} \leq P_1 \leq 650\text{ MW}$$
$$150\text{ MW} \leq P_2 \leq 380\text{ MW}$$
$$100\text{ MW} \leq P_3 \leq 260\text{ MW}$$

A load curve data is recorded in Tables 3.11–3.12.

TABLE 3.11 Load Curve Data

Time (Hours)	Load (MW)
0 ~ 4	1200
4 ~ 8	575
8 ~ 12	500
12 ~ 20	520
20 ~ 22	600
22 ~ 25	900
25 ~ 32	1000

Calculate the total consumed energy, and then, schedule the operation of the units for the individual period (Figure 3.9). The total maximum generating power of the units:

Unit 1 + Unit 2 + Unit 3	= 1290 MW
Unit 1 + Unit 2	= 1030 MW
Unit 1 + Unit 3	= 910 MW
Unit 2 + Unit 3	= 640 MW
Unit 1	= 650 MW
Unit 2	= 380 MW
Unit 3	= 260 MW

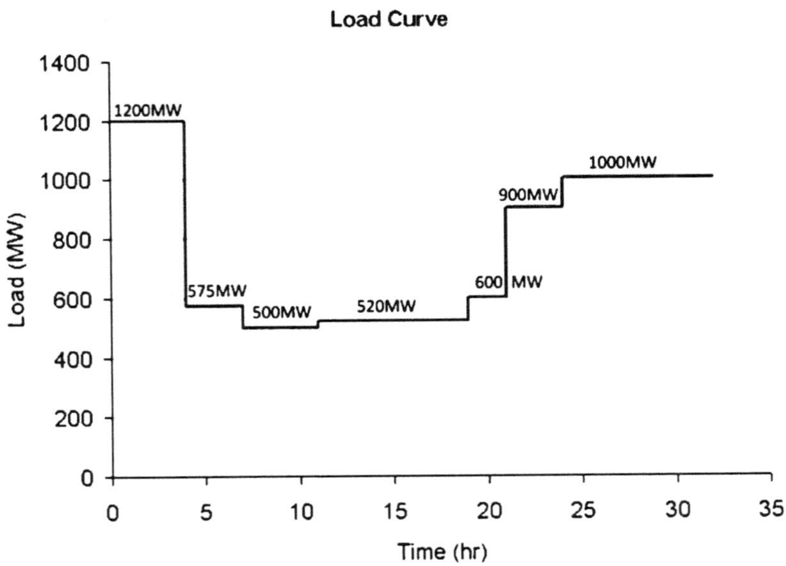

FIGURE 3.9 Load curve for the example (3.11).

The consumed energy = area under the curve = 24160 MW hr.

3.6 PRIORITY LIST METHOD (PLM)

In some cases, we need to operate the system at full capacity, which means that each unit will operate at its full capacity. This is the reason that the priority list method (PLM) is needed. In PLM, units should be on line according to some of the defined characteristics of them, such as maximum demand and maximum capacity. The reason behind the units to be on-line will help to satisfy the summation of the load demand and spinning reserve. The PLM is a fast and simple method that is modified to find unit scheduling. Most likely, the priority list will become and decided based on the fuel consumption at the maximum capacity of each unit (Quan et al., 2015; Delarue et al., 2013).

TABLE 3.12 Three Units Status

Load (MW)	Unit 1	Unit 2	Unit 3
1200	ON	ON	ON
575	ON	OFF	OFF
	OFF	ON	ON
500	ON	OFF	OFF
	OFF	ON	ON
520	ON	OFF	OFF
	OFF	ON	ON
600	ON	OFF	OFF
	OFF	ON	ON
900	ON	OFF	ON
	ON	ON	OFF
1000	ON	ON	OFF
	ON	ON	ON

Average Load = 24160/32 = 755 MW

Example (3.12):

Three units are considered for the present study:

$$H_1 = 510 + 7.2\,P_1 + 0.00142\,P_1^2 \text{ MBTU/hr}$$
$$H_2 = 310 + 7.85\,P_2 + 0.00194\,P_2^2 \text{ MBTU/hr}$$
$$H_3 = 78 + 7.97\,P_3 + 0.00482\,P_3^2 \text{ MBTU/hr}$$

Their limits respectively are:

Unit 1: $200 \text{ MW} \le P_1 \le 500 \text{ MW}$
Unit 2: $150 \text{ MW} \le P_2 \le 300 \text{ MW}$
Unit 3: $100 \text{ MW} \le P_3 \le 150 \text{ MW}$

and

$f_1 = 1.1$ \$/MBTU
$f_2 = 1.0$ \$/MBTU
$f_3 = 1.2$ \$/MBTU

Follow the PLM and organize the unit on the priority. Then find the maximum loading that the units can reach. Finally, what are the best combinations based on the priority?

Solution:

$Hi\ (P_i \max)\ f_i/P_i \max$

TABLE 3.13 Calculations

Unit	$H_i@P_i$ max MBTU/hr	$(H_i@P_{i\,max})f_i = C_i@P_i$ max	$\frac{C_i@P_{i\,max}}{P_{i\,max}}$	Priority
1	4465	4911.5	9.823	2
2	2839.6	2839.6	9.465	1
3	1381.95	1658.34	11.0556	3

TABLE 3.14 Priority List Method Application

Unit(s) Under Operation	Maximum Load (MW)	$\frac{\sum_{i=1}^{n} H_i(P_i max)f_i}{\sum P_i max}$	Priority
2	300	9.4650	1
1	500	9.8230	3
3	150	11.0556	7
2 + 1 + 3	950	9.9050	4
2 + 1	800	9.6890	2
2 + 3	450	9.9950	5
1 + 3	650	10.107	6

TABLE 3.15 Fin Results for PLM

Unit(s) Under Operation	Maximum Load (MW)	$\frac{\sum H_i(P_i max) f_i}{\sum_{i=1}^{n} P_i max}$	Priority
2	300	9.4650	1
2 + 1	800	9.6890	2
1	500	9.8230	3
2 + 1 + 3	950	9.9050	4
2 + 3	450	9.9950	5
1 + 3	650	10.107	6
3	150	11.0556	7

Example (3.13):

Three generating units are considered for:

$$H_1 = 320 + 6.2\,P_1 + 0.001\,P_1^2 \text{ MBTU/hr}$$
$$H_2 = 170 + 6.85\,P_2 + 0.002\,P_2^2 \text{ MBTU/hr}$$
$$H_3 = 100 + 6.97\,P_3 + 0.005\,P_3^2 \text{ MBTU/hr}$$

Their limits respectively are:

Unit 1: $400 \text{ MW} \leq P_1 \leq 1000 \text{ MW}$
Unit 2: $300 \text{ MW} \leq P_2 \leq 600 \text{ MW}$
Unit 3: $200 \text{ MW} \leq P_3 \leq 300 \text{ MW}$

and

$$f_1 = 1.2 \text{ BD/MBTU}$$
$$f_2 = 1.15 \text{ BD/MBTU}$$
$$f_3 = 1.2 \text{ BD/MBTU}$$

Follow the PLM and organize the unit(s) on the priority. Then find the maximum loading that the units can reach. Finally, what are the best combinations based on the PLM?

Solution:

P_1 (max) =	1000	MW	f1 (BD/MBTU)	1.2
P_2 (max) =	600	MW	f2 (BD/MBTU)	1.15
P_3 (max) =	300	MW	f3 (BD/MBTU)	1.2

				Summation	C @ Pmax	[C@Pmax]/Pmax	Generator
H_1 =	320	6200	1000	7520	9024	9.024	1
H_2 =	170	4110	720	5000	5750	9.583333333	2
H_3 =	100	2091	450	2641	3169.2	10.564	3

Unit(s)	Max Load	[C@Pmax]/Pmax	Priority
1	1000	9.024	1
2	600	9.583333333	5
3	300	10.564	7
1+2	1600	9.23375	2
1+3	1300	9.379384615	3
2+3	900	9.910222222	6
1+2+3	1900	9.443789474	4

3.7 UNIT COMMITMENT SOLUTION METHOD

For any system, when we have N number of units operating for a period with M number of intervals, the maximum number of the combination (Ananda et al., 2018; Jayabarathia et al., 2015; Wang et al., 2018; Zhanga et al., 2019) can be formulated as:

$$\text{Max. Combinations} = [2^N - 1^M]$$

Example (3.14):

When we have a power station of 4-number of generators operating for 1 interval of 1 day, find the maximum number of combination(s).

Solution:

$$\text{Maximum number of combination(s)} = [2^N - 1^M]$$

$$= [2^4 - 1^1] = 15$$

Example (3.15):

Consider the previous example with 24 numbers of intervals.

Solution:

$[2^4 - 1^{24}] = 1.683 \times 10^{28}$ combinations

Example (3.16):

Consider a power plant system with 5, 10, 20, and 40 units. Find the maximum number of possible combinations for each system, taking in the consideration 24 hours as a number of intervals.

Solution:

Power Station Number	Number of Units	Maximum Number of Combinations
1	5	6.2×10^{35}
2	10	1.7×10^{72}
3	20	3.1×10^{144}
4	40	9.7×10^{288}

Example (3.17):

Consider five power plants with 4, 8, 18, 26, and 35 units, respectively. Find the maximum number of possible combinations for each plant, taking into consideration 7 days as a number of intervals.

Solution:

Power Plant	No. of Units	$2^{(\text{Number of Units})}$	$2^{(\text{Number of Units})} - 1$	Maximum Number of Combination
1	4	16	15	170859375
2	8	256	255	7.01102E + 16
3	18	262144	262143	8.50683E + 37
4	26	67108864	67108863	6.12998E + 54
5	35	34359738368	34359738367	5.65391E + 73

KEYWORDS

- **priority list method**
- **simple peak valley curve**
- **spinning reserve**
- **unit commitment solution method**

REFERENCES

Ananda, H., Naranga, N., & Dhillon, J. S. Unit commitment considering dual-mode combined heat and power generating units using integrated optimization technique. *Energy Conversion and Management*, **2018**, *171*, 984–1001.

Beigvand, S. D., Abdi, H., & Scala, M. L. A general model for energy hub economic dispatch. *Applied Energy*, **2017**, *190*, 1090–1111.

Boroojeni, K. G., Amini, M. H., Iyengar, S. S., Rahmani, M., & Pardalos, P. M. An economic dispatch algorithm for congestion management of smart power networks. *Energy Systems*, **2017**, *8*(3), 643–667.

Delarue, E., Cattrysse, D., & D'haeseleer, W. Enhanced priority list unit commitment method for power systems with a high share of renewables. *Electric Power Systems Research*, **2013**, *105*, 115–123.

Hirasea, Y., Noroa, O., Nakagawa, H., Yoshimura, E., S. Katsura, S., Abe, K., Sugimoto, K., & Sakimoto, K. Decentralized and interlink-less power interchange among residences in microgrids using virtual synchronous generator control. *Applied Energy*, **2018**, *228*, 2437–2447.

Hreinsson, K., Vrakopoulou, M., & Andersson, G. Stochastic security constrained unit commitment and non-spinning reserve allocation with performance guarantees. *Electrical Power and Energy Systems*, **2015**, *72*, 109–115.

Jayabarathia, R., Jismab, M., & Suyampulingam, A. Unit commitment using embedded systems. *Procedia Technology*, **2015**, *21*, 96–102.

Jin, X., Mu, Y., Jia, H., Wu, J., Jiang, T., & Yu, X. Dynamic economic dispatch of a hybrid energy microgrid considering building based virtual energy storage system. *Applied Energy*, **2017**, *194*, 386–398.

Jordehi, A. R. How to deal with uncertainties in electric power systems? A review. *Renewable and Sustainable Energy Reviews*, **2018**, *96*, 145–155.

McLarty, D., Panossian, N., Jabbari, F., & Traverso, A. Dynamic economic dispatch using complementary quadratic programming. *Energy*, **2019**, *166*, 755–764.

Odetayo, B., MacCormack, J., Rosehart, W. D., Zareipour, H., & Seif, A. R. Review: Integrated planning of natural gas and electric power systems. *Electrical Power and Energy Systems*, **2018**, *103*, 593–602.

Palacio, S. N., Kircher, K. J., & Zhang, K. M. On the feasibility of providing power system spinning reserves from thermal storage. *Energy and Buildings*, **2015**, *104*, 131–138.

Quan, R., Jian, J., & Yang, L. An improved priority list and neighborhood search method for unit commitment. *Electrical Power and Energy Systems*, **2015**, *67*, 278–285.

Wang, J., Guo, M., & Liu, Y. Hydropower unit commitment with nonlinearity decoupled from mixed integer nonlinear problem. *Energy*, **2018**, *150*, 839–846.

Wang, S., Hui, H., Ding, Y., & Zhu, C. Cooperation of demand response and traditional power generations for providing spinning reserve. *Energy* Procedia, **2017**, *142*, 2035–2041.

Zakian, P., & Kaveh, A. Economic dispatch of power systems using an adaptive charged system search algorithm. *Applied Soft Computing Journal*, **2018**, *73*, 607–622.

Zhanga, Y., Hana, X., Yanga, M., Xub, B., Zhaoa, Y., & Zhaic, H. Adaptive robust unit commitment considering distributional uncertainty. *Electrical Power and Energy Systems*, **2019**, *104*, 635–644.

Zimmer, P. D., & Gabet, E. J. Assessing glacial modification of bedrock valleys using a novel approach. *Geomorphology*, **2018**, *318*, 336–347.

CHAPTER 4

Overview of Power System Reliability

4.1 BASIC POWER SYSTEMS RELIABILITY CONCEPTS

North American Electric Reliability Corporation (NERC, 2012, 2018) defines the term reliability of the interconnected bulk-power system in terms of two basic and functional aspects. These aspects are adequacy and operating reliability. In the present text, these two main terms are defined. The bulk power system is a combination of three main parts: generation, transmission, and distribution. The main terms used in the electric industry are the reliable and unreliable system. The performance of the power system is measured by calculating the reliability. The reliability term is a result of an acceptable standard and is the amount wanted. In this case, to meet the reliability of consumers, it requires conditions for the bulk power system. These requirements are to find the plan, design, construct, operate, maintain, and restore of the power system (Kadhema et al., 2017; Billinton and Allan, 1992; Rusin and Wojaczek, 2015).

4.1.1 POWER SYSTEM OPERATIONAL LIMITATION RESTRICTION

A multiplicity of factors must be considered of bulk power systems. This will help in the reliability assessment. A question might be raised. This question concentrates on a number of factors helping in the reliability assessment. These factors are the failure and repair rates of any power equipment and operating practices (Balasubramaniam et al., 2016; Georgescu et al., 2018; Van Stiphout et al., 2017). Other factors are the economic generation scheduling, security controls, projected load variations, emergency controls, and maintenance schedules for the considered system. The system security relates to the system's ability to respond to arising disturbances within the system (Boroujeni et al., 2012).

4.1.2 FUNCTIONAL ZONES AND HIERARCHICAL LEVELS

The functional zones of generation, transmission, and distribution are representing a complete electric power system (Billinton and Allan, 1992). The combination of the three functional zones known as the hierarchical levels. Electric power system reliability evaluation consists of three main levels. These levels are known as the hierarchical level. The first level called hierarchical level one (HLI), where the second called hierarchical level two (HLII), and the third level is hierarchical level three (HLIII). HLI is concerned with Generation system evaluation while the second level, which is HLII considers composite generation and transmission system evaluation. Finally, HLIII evaluates distribution system reliability. These three hierarchical levels are shown in Figure 4.1 (Moghadam and Abdi, 2013; Vrana and Johansson, 2011; Billinton and Allan, 1996).

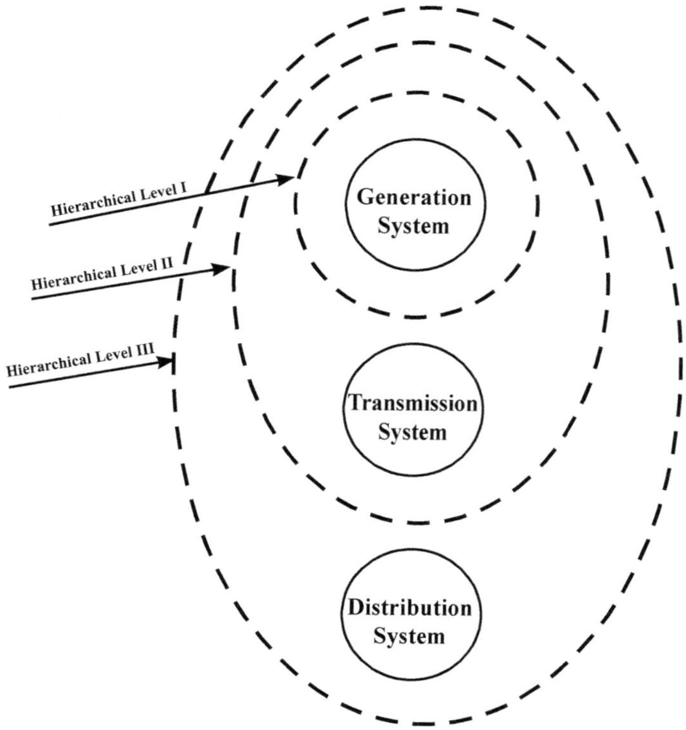

FIGURE 4.1 Three hierarchical levels.
Source: Based on Billinton and Allan, 1992.

Based on the power system structure and the concept of the three hierarchical levels, the following subsystems are defined:

i. **Power Generation Stations:** The analysis of each power station or even each generating-unit is carried out separately. The analysis helps with establishing equivalent components. The system has some reliability indices. The components established inputs to both (HLI) and (HLII).
ii. **Power Generation Capacity:** The process of plan a generating-sources (HLI) to meet the loads requirement as economically as possible and help with finding the reliability of generating capacity.
iii. **International System:** Finding the model for each power system and the tie lines between them, where the networks between them are not considered (HLI).
iv. **Made Up of Generation/Transmission:** Bulk of transmission presenting electrical network. This bulk is evaluated based on its ability and quality. This hierarchy is categorized in (HLII).
v. **Distribution Networks:** The electrical network is fed from bulk supply points. The lead point indices assessed in the (HLII), which can be used as input values if the overall (HLIII) indices are needed.
vi. **Electrical Substations:** When any substation is analyzed separately. It means that the indices of which can be used to measure the performance of the substation itself or evaluating the reliability of transmission lines (in HLII) or distribution (in HLIII) system.

4.1.3 ELECTRIC POWER SYSTEMS RELIABILITY ASSESSMENT PROCEDURE

The adequacy and operating reliability is a procedure to measure the ability of the power system to supply electric power. The probabilistic techniques such as analytical technique and Monte Carlo simulation are used as techniques for reliability assessment of composite power system (Tripathi and Sisodia, 2014). At the same time, the fault tree analysis technique is used for assessments of the power system.

4.1.4 ADEQUACY AND OPERATING RELIABILITY

The term adequacy (NERC, 2012, 2018; AEMO, 2011) is defined as a measure of the ability of the power system to supply the electric power and energy cumulative requirements of the customers within components ratings and voltage limits, taking into consideration planned and unplanned outages of system components. The adequacy term is the measurement of the capability of the power system to supply the load in all the steady states. The adequacy term relates to the existence of sufficient facilities within the power system to satisfy the consumer load demand at all times. The term that considers any region's energy and maximum demand expectations, and limitations to the amount of energy scheduled generating units can supply due to a range of factors including the deficiency in fuel and cooling the water restrictions. The performance state of any system is known as an adequate level of reliability. In this case, the design, planning, and operation of the bulk electric system will satisfy the objectives and performance of the system's reliability.

The power reliability operating (NERC, 2018, 2019; AEMO, 2011) defined and assessed in different states. The study also provides an opportunity for the preparations to ensure reliability. In the present study, it can be defined as:

i. The normal state, where in this case, all system variables are in the normal range. This means that no equipment is being overloaded. The system can operate in a secure manner and is able to withstand a contingency without violating any of the constraints.
ii. The alert (warning) state, security level falls below a certain limit of adequacy. Due to the change of weather, the possibility of a disturbance increases. It is assumed that all system variables are still within the acceptable range, and all constraints are satisfied.
iii. The emergency state, by emergency actions, any electric power system can be restored back to an alert state. Some examples for the emergency state are the fault clearing, excitation control, generation tripping, generation runback, and load shedding.
iv. The state that can be called in extremis state. This type of state is recognized when the system goes to in extremis state, which is that the emergency measures are not applied or are ineffective. In this state, there will be a possibility of a shutdown of the major part of the system. To save the system, control actions such as load

shedding and controlled separation are required. The requirements of such actions could save much of the system from a possible blackout.

v. Restorative state, this state highlights on how to improve and repair the system. In this case, the system could either pass through the normal state directly or through the alert state. It means the state depending on the conditions.

4.2 BATH-TUB CURVE

Any system should pass through different periods. Figure 4.2 shows the generating unit passing through three periods (Billinton and Allan, 1992).

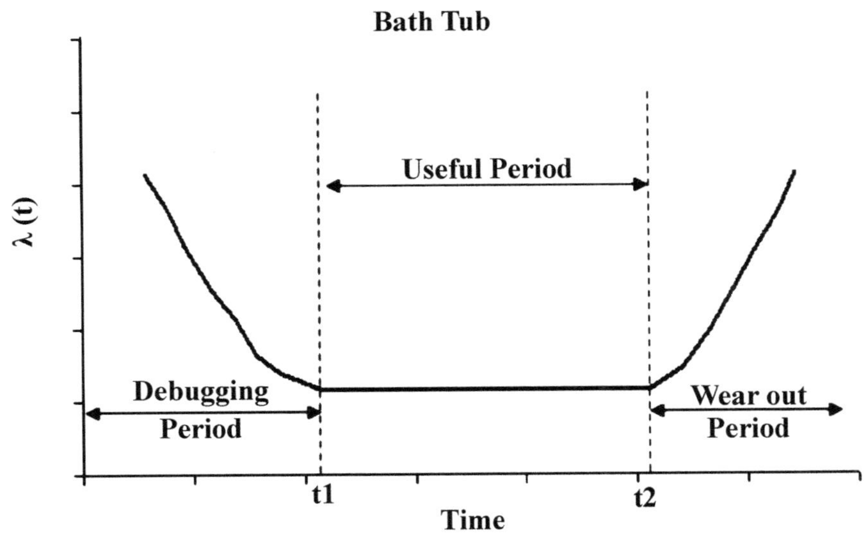

FIGURE 4.2 Bath-tub curve.

1. **Debugging Period:** During this period, the system (or unit) is new, where the risk is high.
2. **Useful Period (Useful Life):** This period considered as the optimal and best period of the unit (equipment).
3. **Wear Out Period:** This period is the worst period of the unit, where we expect that many changes in the components needed.

4.3 TWO-STATE MODEL

Consider a generator passing through two-states (Figure 4.3), the first-state is considered as an operated state and the second-state as a fail state (Billinton and Allan, 1992).

State 1: P_1 (t): probability number (1).
State 2: P_2 (t): probability number (2).

$$P_1(t) + P_2(t) = 1$$

When we said the summation of both probabilities (Figure 4.4) is 1, it means that it is 100%.

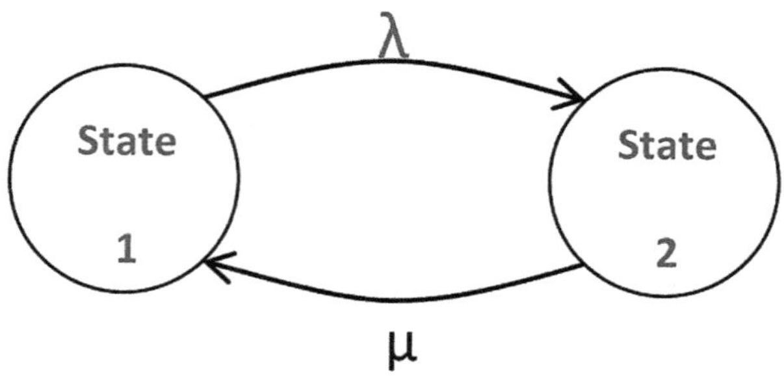

FIGURE 4.3 Two-state model.
Source: Based on Billinton and Allan, 1992.

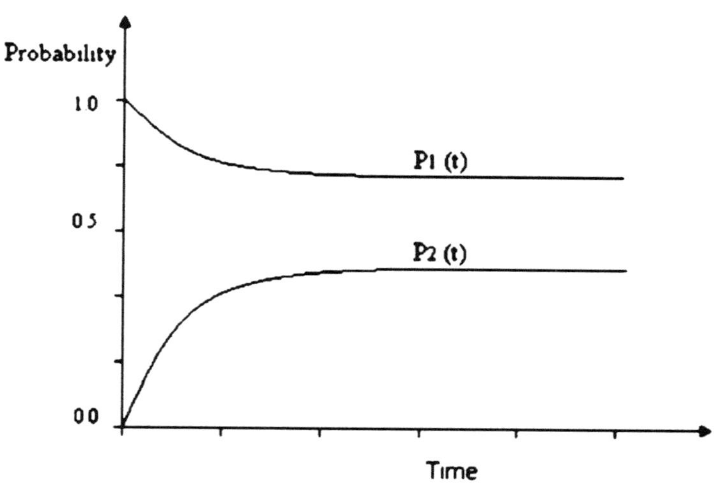

FIGURE 4.4 Two-state probabilities.

Based on the given curve:

The initial probabilities are P_1 (t = 0) =1 and P_2 (t = 0) = 0.

Initial probabilities:

P_1 (t = 0) = 1

P_2 (t = 0) = 0

The initial probabilities are the starting values (Figure 4.5) of the probabilities (which means at t = 0). The final results are recorded in Table 4.1.

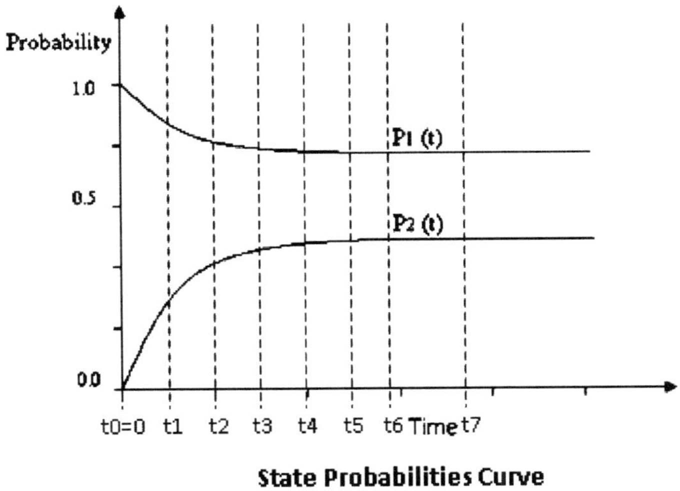

FIGURE 4.5 Second two-state model.

TABLE 4.1 Two-State Model Results

t_i	t_0	t_1	t_2	t_3	t_4	t_5	t_6
$P_1(t_i)$	$P_1(t_0)$	$P_1(t_1)$	$P_1(t_2)$	$P_1(t_3)$	$P_1(t_4)$	$P_1(t_5)$	$P_1(t_6)$
$P_2(t_i)$	$P_2(t_0)$	$P_2(t_1)$	$P_2(t_2)$	$P_2(t_3)$	$P_2(t_4)$	$P_2(t_5)$	$P_2(t_6)$
$\sum_{j=1}^{2} P_j(t_i)$	1	1	1	1	1	1	1

4.4 THREE-STATE MODEL

Consider the three-model state-diagram (Figure 4.6) (Groß et al., 2018).

- **State 1:** It means that the system is with the full capacity.
- **State 2:** It means that the system is working with the partial of its capacity (called a derated state).
- **State 3:** It means that the system is in the fail state.

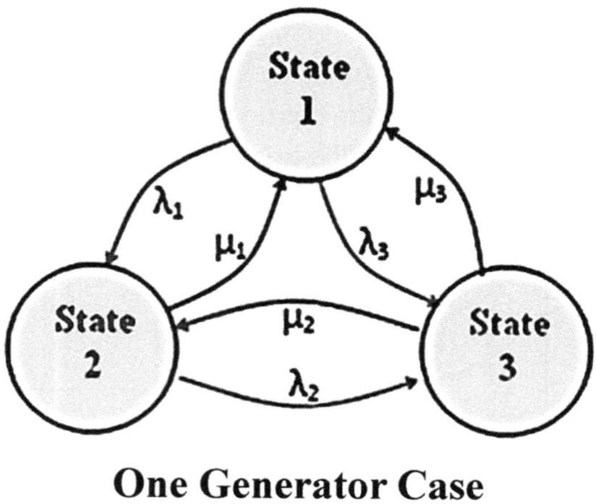

FIGURE 4.6 Three-state model.
Source: Based on Billinton and Allan, 1992.

As more clarification, the derated state might be defined as a technique used in the electric power and electronic devices. At the same time, the derating as a term used when a device or component rated less than the maximum capability of the device or component. In another term, it includes operation below the maximum power rating. The expected excess of available generation capacity over electric power demand is known as the derated capacity margin. The installed capacity is called available generation capacity. This available capacity is expected to be accessible electric power in reasonable operational timelines. The available generation capacity will also take into account any occurring at irregular intervals of electric power demand. Sometimes, the simple capacity margin is used in a way it indicates that the system is less reliable. At the same time, the derated capacity margin statement is used. This is because uncontinuous plant generation depends on weather conditions, and thus, these types of plants provide uncertain levels of generation capacity. At non-peak demand times, the electricity system could also be under stress sometimes. This means that the electric system might be under maintenance outages. Traditionally, the derated capacity margin is measured relative to peak demand.

$$P_1(t) + P_2(t) + P_3(t) = 1$$

When we have two identical generators (Figure 4.7), where:

- **State 1:** Means that both generators are working with their full capacity.
- **State 2:** Means that one is operating and the other is in the fail-mode.
- **State 3:** Means that both are in the fail-mode.

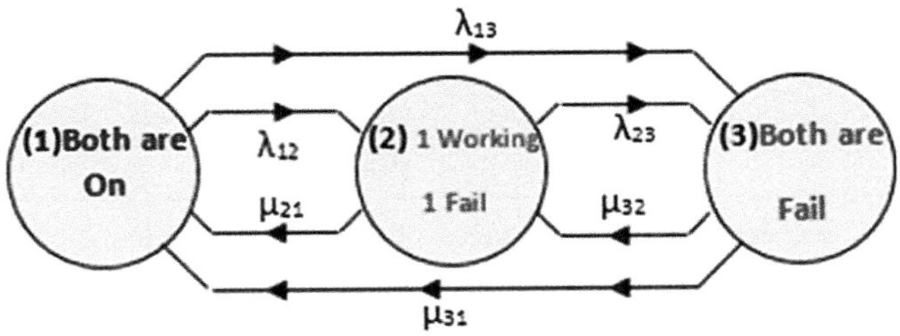

FIGURE 4.7 Two generator case.

Q: What are the initial probabilities of the above curve?
A: The initial probabilities (Figure 4.8) are:

$$P_1(t = 0) = 1,\ P_2(t = 0) = P_3(t = 0) = 0.$$

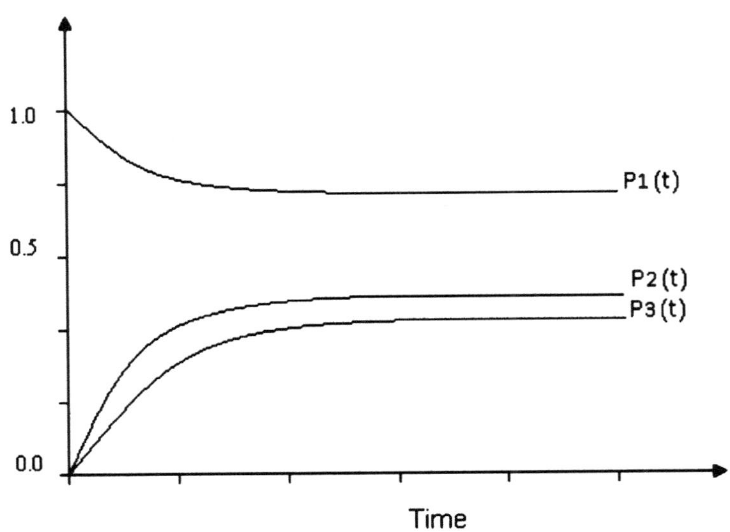

FIGURE 4.8 Three-state results.

In general, at any time on the curve (2 – State, 3 – State or any number of states) the summation of the probabilities will equal to unity. Some basic definitions and terms are illustrated in Table 4.2.

TABLE 4.2 Basic Definitions and Terms

State	Opposite State
Up	Down
Operating	Fail
Full Capacity	Zero Capacity
High	Low
Commit	Uncommit
Success	Fail

Probabilities of Success = No. of Success/(No. of Success + No. of Fail).
Probabilities of fail = No. of Fail/(No. of Success + No. of Fail).

4.5 MEAN TIME INDICES

There are number indices in the reliability studies (Subburaj, 2015). Some of these are the mean time to repair (MTTR), mean time to fail (MTTF) and the mean time between fail (MTBF). These terms can be defined based on figure (VV) as follow:

$$MTTR = \frac{1}{\mu}$$

In other form, the MTTR is expressed in the following form:

$$MTTR = \frac{Total\ Maintenance\ Time}{Number\ of\ Repair\left(Number\ of\ Maintenance\ Actions\right)}$$

The MTTR is called mean downtime (MDT) or repair time of systems.

In case that the equipment needs too long time to repair, it means the equipment needs a replacement and scraped.

$$MTTF = \frac{1}{\lambda}$$

The MTTR expression is defined as the total maintenance time divided by the total number of maintenance actions over a specific period. In other words, it is a basic technical measure of the maintainability of equipment and repairable parts. Also, the term MTTF (Çekyay and Özekici, 2015) is defined as to run controlled tests for component to see how reliable a component is expected to be. Sometimes report the results coming out of the carried-out test. This is a good indication of the reliability of a component, as long as these tests are reasonably accurate. The MTTF is a basic measure of reliability for non-repairable components. The MTTF is calculated as the total time of operation divided by the total number of components. It has the following expression:

$$MTTF = \frac{Total\ Hours\ of\ Operation}{Total\ Number\ of\ Components}$$

As an example, if it is considered that three components are considered. The first component is failed after 20 years, the second one failed after 14 years, while the third failed at 17 years. Therefore, the MTTF is calculated as:

$$MTTF = \frac{20+14+17}{3} = 17\,Years$$

This means the component will need to be replaced on average of 17 years.

Example (4.1):

A transformer operates during its normal life period. It is known that 15% of these transformers fail in a 300-hours period. Find their MTTF and the time periods during 25%, 40%, and 65% are likely to fail.

Solution:

From the example, it is clear that 15% of the available Transformers are failed in a time of 300 hours.

$$\because \quad R(t) = e^{-\lambda t}$$

and

$$R(t) + Q(t) = 1$$

$$\therefore \quad R(t) + 0.15 = 1$$

$$R(t) = 1 - 0.15 = 0.85$$

$$0.85 = e^{-\lambda(300)}$$

Taking the (Ln) for both sides:

$$\ln(0.85) = -300\,\lambda$$

$$-0.162519 = -300\,\lambda$$

$$\therefore \quad \lambda = 0.00054173 \text{ per hour}$$

$$\because \quad MTTF = \frac{1}{\lambda}$$

$$\therefore \quad MTTF = \frac{1}{0.00054173} = 1845.938 \text{ hours}$$

For 25% to fail, to find the time-period during 25%:

$$0.75 = e^{-(0.00054173)t}$$

$$- \quad 0.288 = -\,(0.00054173)\,t$$

$$\therefore \quad t = 531.043 \text{ hours}$$

For 40% to fail, to find the time-period during 40%:

$$0.6 = e^{-(0.00054173)t}$$

$$- \quad 0.51083 = -\,(0.00054173)\,t$$

$$\therefore \quad t = 942.952 \text{ hours}$$

For 65% to fail, to find the time-period during 65%:

$$0.35 = e^{-(0.00054173)t}$$

$$- \quad 1.05 = -\,(0.00054173)\,t$$

$$\therefore \quad t = 1937.907 \text{ hours}$$

The predicted elapsed time between inherent failures of electrical power equipment is called MTBF during the operation of the normal system (Figure 4.9). The MTBF is defined as the average time that a system will run between failures as well. In other words, it is the average between successive failures. The calculation of the MTBF can be carried out as the arithmetic mean (average) time between failures of the electrical system or any other system. MTBF is commonly used for both repairable and non-repairable components. MTBF is another indicator of reliability.

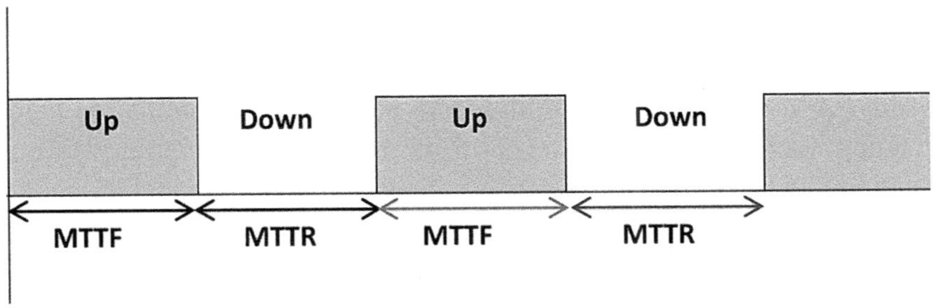

FIGURE 4.9 MTTF and MTTR.
Source: Based on Billinton and Allan, 1992.

The term can be presented as follows:

$$MTBF = \frac{T}{n}$$

where T is the total run time of the system, and (n) is the number of failures during the considered time of that system. The total run time (T) of the system is the summation of operational times. In the other form, the MTBF is expressed in the following form:

$$MTBF = \frac{\sum (Beginning\ of\ Downtime - Start\ of\ Raisedtime)}{Number\ of\ Failures}$$

MTBF used to project how a single component is to fail within a certain period of time. To quantify the reliability of equipment or system is a great way of that component or system. At the same time, the MTBF can be expressed as:

$$MTBF = \frac{1}{\lambda_1 + \lambda_2 + \lambda_3 + \ldots + \lambda_n}$$

where: λ_i is the failure rate of the i – th component.

This means that the MTBF is a reciprocal of the failure rate.

Example (4.2):

An electric power network is running for 40 weeks before failure, then 35 weeks and then 49 weeks. Calculate the MTBF of the network (4.10).

Solution:

From the given data, the total run time would be the summation of the running weeks, which is 124 weeks. This means that the failures (n) would be three times (Figure 4.10).

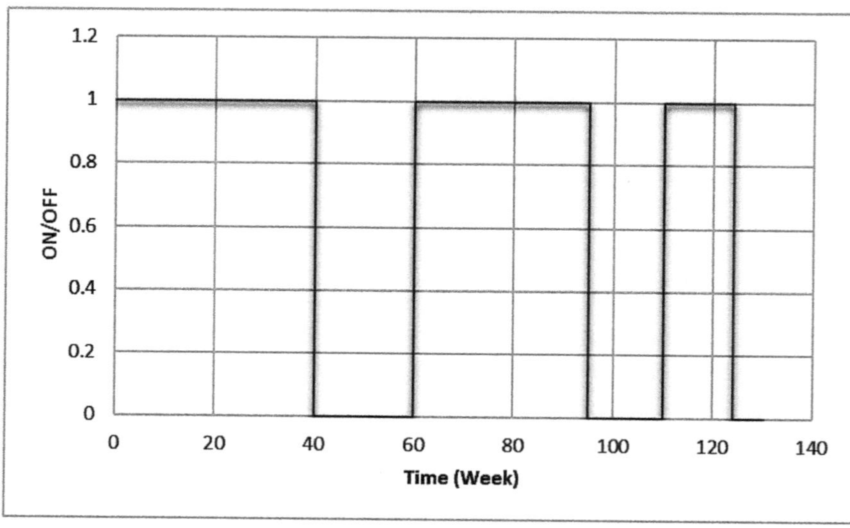

FIGURE 4.10 Electric power network running.

This would the running time divide by the number of failures (n) to find the MTBF, which becomes:

$$MTBF = \frac{T}{n}$$

$$MTBF = \frac{124}{3} = 41.33 \text{ weeks}$$

Therefore, the MTBF is 41.33 weeks of operation before the failure occurred.

KEYWORDS

- **MTBF**
- **MTTF**
- **MTTR**
- **meantime to fail**
- **meantime to repair**
- **North American Electric Reliability Corporation**

REFERENCES

AEMO (Australian Energy Market Operator). *Power System Adequacy for National Electricity Market*. Australia, **2011**.

Balasubramaniam, K., Abdlrahem, A., Hadidi, R., & Makram, E. B. Balanced, non-contiguous partitioning of power systems considering operational constraints. *Electric Power Systems Research*, **2016**, *140*, 456–463.

Billinton, R., & Allan, R. N. *Reliability Evaluation of Engineering Systems: Concepts and Techniques* (2nd edn.). Pitman: New York, **1992**.

Billinton, R., & Allan, R. N. *Reliability Evaluation of Power Systems* (2nd edn.). Plenum Press: New York, **1996**.

Boroujeni, H. F., Eghtedari, M., Abdollahi, M., & Behzadipour, E. Calculation of generation system reliability index: Loss of load probability. *Life Science Journal*, **2012**, *9*(4), 4903–4908.

Çekyay, B., & Özekici, S., Reliability MTTF and steady-state availability analysis of systems with exponential lifetimes. *International Journal of Applied Mathematical Modeling*, **2015**, *39*, 284–296.

Chen, F., Li, F., Feng, W., Wei, Z., Cui, H., & Liu, H. Reliability assessment method of composite power system with wind farms and its application in capacity credit evaluation of wind farms. *Electric Power Systems Research*, **2019**, *166*, 73–82.

Feng, D., Lin, S., Sun, X., & He, Z. Reliability assessment for traction power supply system based on quantification of margins and uncertainties. *Microelectronics Reliability*, **2018**, *88–90*, 1195–1200.

Georgescu, I., Godjevac, M., & Visser, K. Efficiency constraints of energy storage for onboard power systems. *Ocean Engineering*, **2018**, *162*, 239–247.

Groß, D., Arghir, C., & Dörfler, F. On the steady-state behavior of a nonlinear power system model. *Automatica*, **2018**, *90*, 248–254.

Kadhema, A. A., Abdul Wahab, N. I., Aris, I., Jasni, J., & Abdalla, A. N. Computational techniques for assessing the reliability and sustainability of electrical power systems: A review. *Renewable and Sustainable Energy Reviews*, **2017**, *80*, 1175–1186.

Moghadam, S., & Abdi, H. Reliability analysis of large scale wind power plants over a year. *Technical Journal of Engineering and Applied Sciences*, **2013**, *3*(24), 3541–3546.

NERC (North American Electric Reliability Corporation). *Probabilistic Adequacy and Measures: Technical Report Final*. Reliability Accountability: Atlanta, **2018**.

NERC (North American Electric Reliability Corporation), *NERC's 2019–2020 Winter Reliability Assessment Report,* November **2019**.

NERC (North American Electric Reliability Corporation). *Reliability Assessment Guidebook*. Version 3.1. Reliability Accountability: Atlanta, **2012**.

Rusin, A., & Wojaczek, A. Trends of changes in the power generation system structure and their impact on the system reliability. *Energy*, **2015**, *92*, 128–134.

Subburaj, R. *Software Reliability Engineering* (1st edn.). McGraw Hill (India) Private Limited, New Delhi, **2015**.

Tripathi, A., & Sisodia, S. Comparative study of reliability assessment techniques for composite power system planning & applications. *International Journal of Engineering Research and Applications (IJERA)*. **2014**, 8–13.

Van Stiphout, A., Brijs, T., Belmans, R., & Deconinck, G. Quantifying the importance of power system operation constraints in power system planning models: A case study for electricity storage. *Journal of Energy Storage*, **2017**, *13*, 344–358.

Vrana, T. K., & Johansson, E. *Overview of Power System Reliability Assessment Techniques*, Cigre, **2011**.

CHAPTER 5

Power Systems Reliability

5.1 INDIVIDUAL AND CUMULATIVE PROBABILITIES AND FREQUENCY

In the long run, the probability of an event is its relative frequency (i.e., expected proportion). As an example, in case that an event occurs *x* times out of n number of times, then its probability will converge on *x* divided by n number of times occurred. For example, if we throw a dice, it is expected to see one-sixth if thrown once as mentioned that number 5 is coming out. However, as the number of observations n increases, the observed frequency becomes a more reliable reflection of the probability. Otherwise, an event is not being a reliable reflection of its probability (Billinton and Allan, 1992; Pérez-Ràfols and Almqvist, 2019).

5.2 PROBABILITY, FREQUENCY, AND PROBABILITY DISTRIBUTION

The term probability is defined as an event to occur when an experiment is carried out. The probability can only be calculated for random variables. The probability distribution of a discrete random variable (e.g., electric load) is presented by a probability histogram as the relationship between the electric load and its probability of occurrence. In the same way, the probability graph is the height of each rectangle of the individual probability histogram, which is equal to the probability that the random variable takes on the value, which corresponds to the mid-point of the base (Pérez-Ràfols and Almqvist, 2019; He and Zheng, 2018).

Frequency is defined as the repetition of data in a data set. Relative frequency is calculated as the frequency of certain data in a given data set divided by the total number of frequencies. The total number of frequencies and the total number of data may or may not be the same. The relative frequency sometimes becomes the same as a theoretical probability, but they are not the same every time.

Example (5.1):

A small power plant recorded a peak loads in MW over 20 day's period: 5.97, 5.99, 6.01, 6.00, 6.00, 5.99, 5.98, 6.02, 5.97, 5.98, 5.99, 5.99, 6.00, 6.02, 6.01, 6.02, 6.00, 5.99, 5.98, and 6.00.

Find how frequent each peak load occurred, and then calculate the cumulative probabilities of the peak loads. Finally, find the expectation of the loads events.

Solution: (Table 5.1)

TABLE 5.1 Data and Results for Example (5.1)

Load (MW)	Frequency	Individual Probability	Cumulative Probability	Expectation of the Load Events
5.97	2	2/20 = 0.10	0.10	(5.97)(0.10) = 0.597
5.98	3	3/20 = 0.15	0.10 + 0.15 = 0.25	(5.98)(0.15) = 0.897
5.99	5	5/20 = 0.25	0.25 + 0.25 = 0.50	(5.99)(0.25) = 1.4975
6.00	5	5/20 = 0.25	0.50 + 0.25 = 0.75	(6.00)(0.25) = 1.500
6.01	2	2/20 = 0.10	0.75 + 0.10 = 0.85	(6.01)(0.10) = 0.601
6.02	3	3/20 = 0.15	0.85 + 0.15 = 1.00	(6.02)(0.15) = 0.903
	Σ=20 Days	Σ= 1.00		

a. Sketch the frequency histogram to the peak load (Figure 5.1).

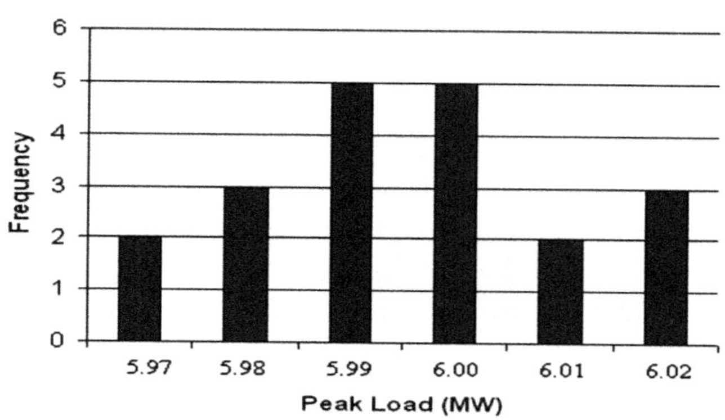

FIGURE 5.1 Frequency histogram to the peak load.

b. Sketch the individual probabilities to the peak load (Figure 5.2).

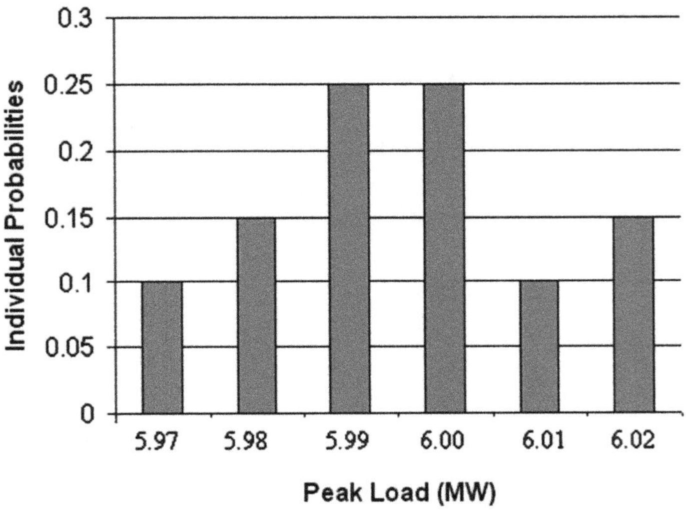

FIGURE 5.2 Individual probabilities to the peak load.

c. Sketch the cumulative probabilities histogram (Figure 5.3).

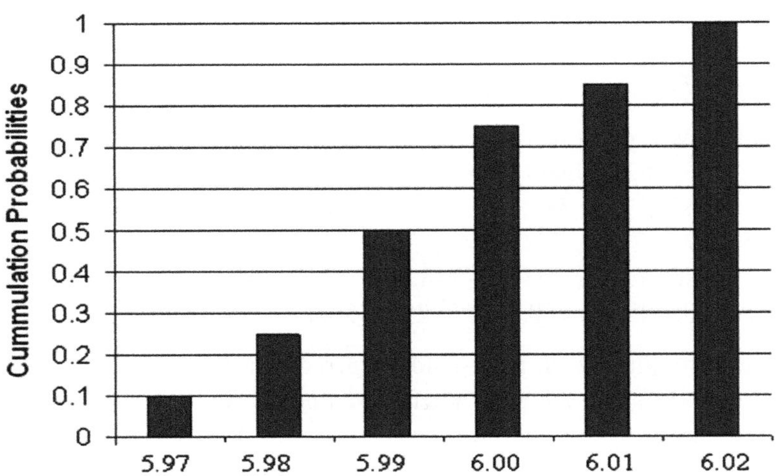

FIGURE 5.3 Cumulative probabilities histogram.

Example (5.2):

A load point has a number of peak loads in MW over a period of 25 days. These peak loads are 300, 290, 320, 310, 285, 290, 315, 285, 320, 310, 320, 315, 320, 315, 300, 285, 300, 285, 290, 290, 320, 290, 285, 315, and 290. Calculate the following:

a. The probabilities of occurrences of each peak load.
b. The cumulative probabilities.
c. The expectation of each load occurrence.
d. The variance, and
e. The standard deviation.

Solution:

Based on the available data given in the example and as a first step, each load has to be counted how many times occurred (Frequency, days). The results as requirements to find and solution are given in Table 5.2.

The variance is calculated using the following equation:

$$V(X) = \text{Probability} \times [X_i - E(X_i)]^2$$

where: X_i is the Peak Load (in MW); $E(X_i)$ is the Expectation of Peak Load Events= Peak Load × Probability.

The histograms of the results are shown in Figure 5.4 to Figure 5.8). Finally, the Standard Deviation (σ) = $\sqrt{V(X)}$ = 247.47512.

Example (5.3):

A small power plant recorded over 50 days a peak loads in MW: 737.5, 746.25, 750, 752.5, 748.75, 748.75, 750, 737.5, 752.5, 747.5, 750, 748.75, 747.5, 748.75, 750, 737.5, 751.25, 748.75, 737.5, 750, 746.25, 751.25, 748.75, 747.5, 752.5, 752.5, 746.25, 748.75, 746.25, 747.5, 748.75, 752.5, 747.5, 752.5, 752.5, 751.25, 748.75, 747.5, 752.5, 752.5, 750, 748.75, 748.75, 750, 748.75, 750, 752.5, 750, 750, and 751.25.

Find how frequent each peak load occurred. Then calculate the cumulative probabilities of the peak loads and find the expectation of the load events. Finally, sketch the frequency histogram for the peak loads, individual probabilities, and cumulative probabilities.

TABLE 5.2 Results for Example (5.2)

X_i = Peak Load (MW)	Frequency Probability (days)	Individual Probability	Cumulative Probabilities	$E(X_i)$ = Expectation of Peak Load Events = Peak Load × Probability	$X_i - E(X_i)$	$[X_i - E(X_i)]^2$	Variance, V(X)
285	5	0.2	0.2	57	228	51984	10396.8
290	6	0.24	0.44	69.6	220.4	48576.16	11658.28
300	3	0.12	0.56	36	264	69696	8363.52
310	2	0.08	0.64	24.8	285.2	81339.04	6507.123
315	4	0.16	0.8	50.2	264.6	70013.16	11202.11
320	5	0.2	1	64	256	65536	13107.2
	25						61235.03

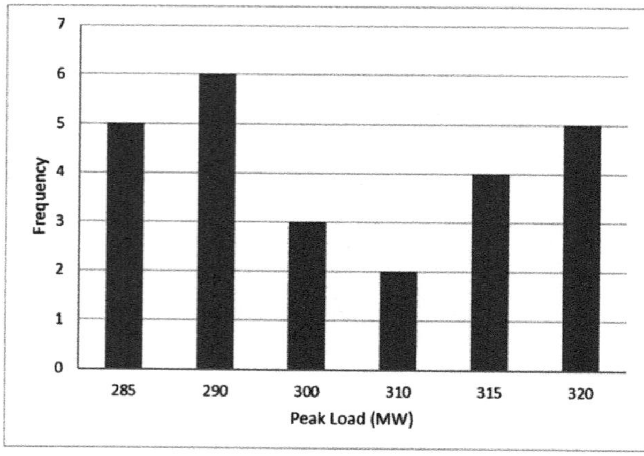

FIGURE 5.4 The frequency vs. peak load.

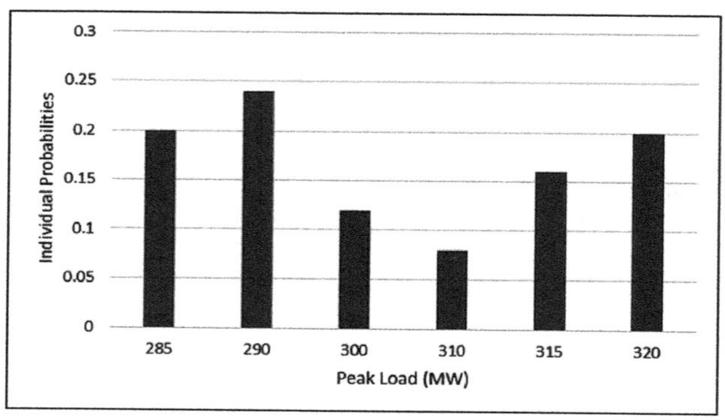

FIGURE 5.5 The individual probabilities vs. peak load.

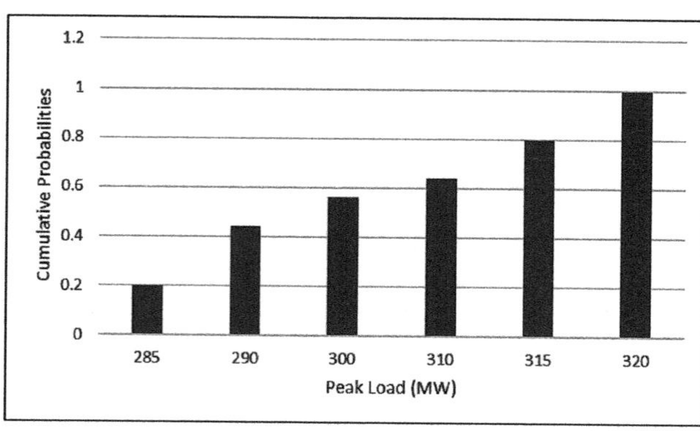

FIGURE 5.6 The cumulative probabilities vs. peak load.

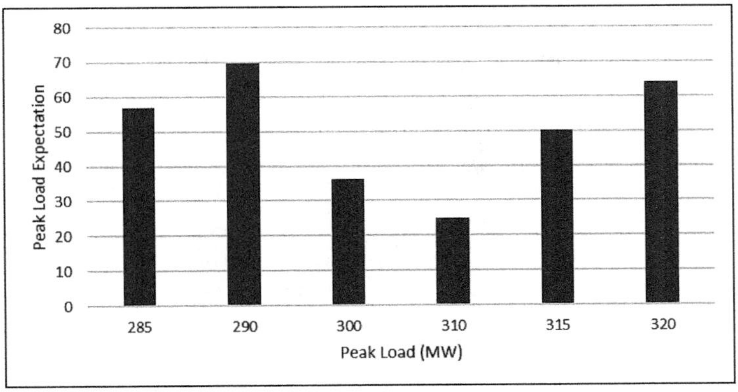

FIGURE 5.7 The peak load expectation vs. peak load.

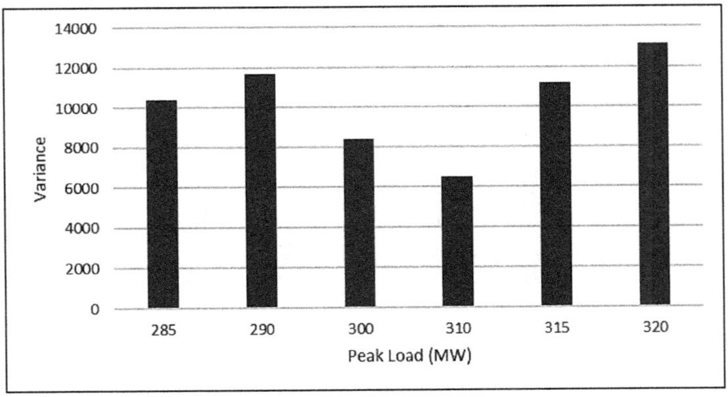

FIGURE 5.8 The variance vs. peak load.

Solution: (Table 5.3)

TABLE 5.3 Results, Example (5.3)

Load (MW)	Frequency	Individual Probability	Cumulative Probability	Expectation of Load Events (MW)
737.5	4	0.08	1	59
746.25	4	0.08	0.92	59.7
747.5	6	0.12	0.84	89.7
748.75	12	0.24	0.72	179.7
750	10	0.2	0.48	150
751.25	4	0.08	0.28	60.1
752.5	10	0.2	0.2	150.5

Figures 5.9–5.11 illustrates the results.

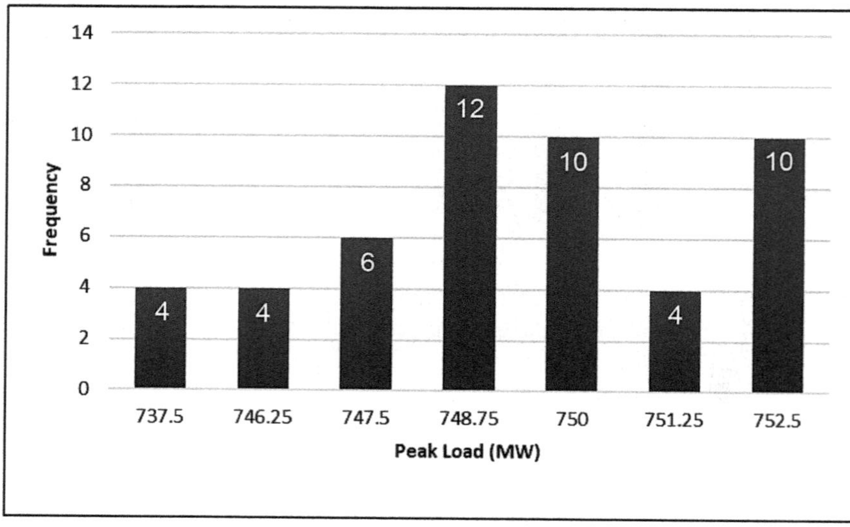

FIGURE 5.9 Frequency vs. peak loads (MW).

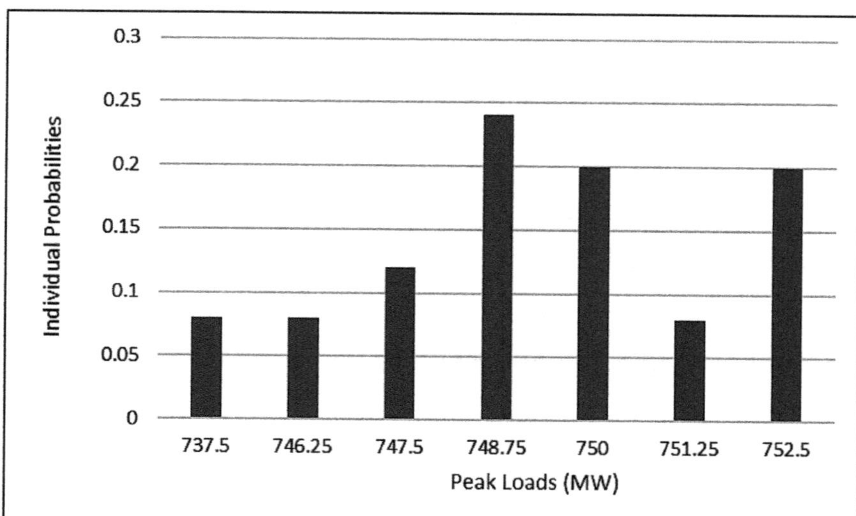

FIGURE 5.10 **(See color insert.)** Individual probabilities vs. peak loads (MW).

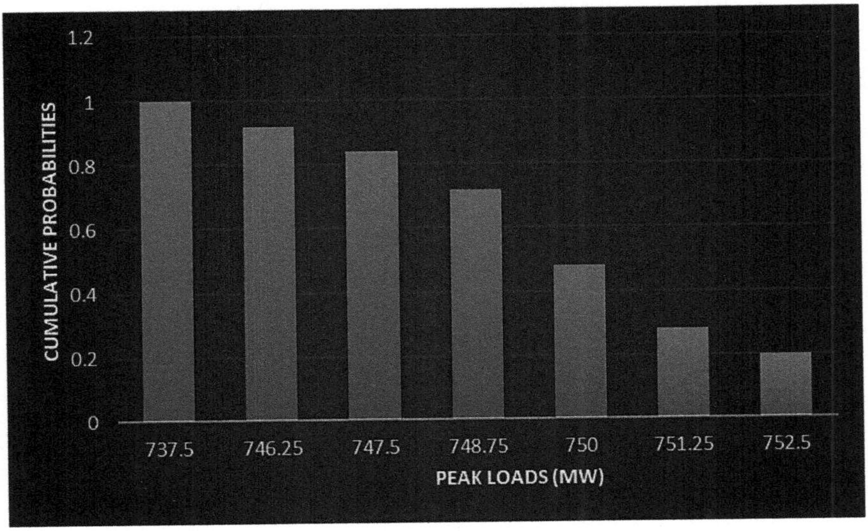

FIGURE 5.11 (See color insert.) Cumulative probabilities vs. peak loads (MW).

5.2.1 SYSTEM RELIABILITY

Reliability of a power system pertains to its ability to satisfy its load demand under the specified operating conditions and policies (Al-Shaalan, 2018; Davidov and Pantos, 2019; Feng et al., 2018).

Example (5.4):

Three systems given below are planned to combine them together. Use the recursive technique when applicable to find their individual probabilities as an overall system (the three systems becoming one) and show the capacities OUT and IN.

System # 1: 3 × 100 MW, each unit with FOR = 0.02 failure/year

System # 2: 2 × 150 MW, each unit with FOR = 0.05 failure/year

System # 3: 2 × 200 MW, each unit with FOR = 0.03 failure/year

Solution: (Table 5.4)

TABLE 5.4 Results, Example (5.4)

System 1	System 2	System 3	SUM *MW*	Individual Probability *PS1*PS2*PS3*	Cumulative Probability	Individual Probability Sum
0.941192	0.9025	0.9409	1000	0.799224716	1	0.7992247
0.057624	0.9025	0.9409	900	0.048932125	0.058808	0.0489321
0.941192	0.095	0.9409	850	0.084128918	0.15057422	0.0841289
0.941192	0.9025	0.0582	800	0.04943658	0.200775284	
0.001176	0.9025	0.9409	800	0.000998615	0.001184	0.0504352
0.057624	0.095	0.9409	750	0.00515075	0.00680234	0.0051508
0.941192	0.0025	0.9409	700	0.002213919	0.06116098	
0.057624	0.9025	0.0582	700	0.003026729	0.009875875	
0.000008	0.9025	0.9409	700	6.7933E–06	8E–06	0.0052474
0.941192	0.095	0.0582	650	0.005203851	0.066445302	
0.001176	0.095	0.9409	650	0.000105117	0.00012266	0.005309
0.941192	0.9025	0.0009	600	0.000764483	0.151338703	
0.057624	0.0025	0.9409	600	0.000135546	0.00132806	
0.001176	0.9025	0.0582	600	6.177E–05	0.000185385	0.0009618
0.057624	0.095	0.0582	550	0.000318603	0.00165159	
0.000008	0.095	0.9409	550	7.15084E–07	7.8E–07	0.0003193
0.941192	0.0025	0.0582	500	0.000136943	0.058947061	
0.057624	0.9025	0.0009	500	4.68051E–05	0.006849145	
0.001176	0.0025	0.9409	500	2.76625E–06	1.094E–05	
0.000008	0.9025	0.0582	500	4.20204E–07	1.2067E–06	0.0001869
0.941192	0.095	0.0009	450	8.04719E–05	0.061241452	
0.001176	0.095	0.0582	450	6.5021E–06	1.75427E–05	8.697E–05
0.057624	0.0025	0.0582	400	8.38429E–06	0.001192514	
0.001176	0.9025	0.0009	400	9.55206E–07	0.000123615	
0.000008	0.0025	0.9409	400	1.8818E–08	2E–08	9.358E–06
0.057624	0.095	0.0009	350	4.92685E–06	0.001332987	
0.000008	0.095	0.0582	350	4.4232E–08	6.4916E–08	4.971E–06
0.941192	0.0025	0.0009	300	2.11768E–06	0.058810118	
0.001176	0.0025	0.0582	300	1.71108E–07	8.17375E–06	
0.000008	0.9025	0.0009	300	6.498E–09	7.86498E–07	2.295E–06
0.001176	0.095	0.0009	250	1.00548E–07	1.10405E–05	1.005E–07
0.057624	0.0025	0.0009	200	1.29654E–07	0.00118413	
0.000008	0.0025	0.0582	200	1.164E–09	1.182E–09	1.308E–07
0.000008	0.095	0.0009	150	6.84E–10	2.0684E–08	6.84E–10
0.001176	0.0025	0.0009	100	2.646E–09	8.00265E–06	2.646E–09
0.000008	0.0025	0.0009	0	1.8E–11	1.80001E–11	1.8E–11

Example (5.5):

Consider two generating systems. The first system is (3 × 23MW) units, and the second system is (2 × 12MW) units, where the FOR-for each unit of both systems is a 0.04 f/year. Calculate the individual probabilities for each system. Then, combine both systems as one and find the individual probabilities. In case that the number of states found after combining both systems need to be reduced to 11 states model with 10 MW interval. Find for the 11 – state model the individual probabilities. Consider each answer at least with six decimal points and show in all cases the capacity OUT and capacity IN.

Solution: (Table 5.5)

System 1								
	FOR=	***0.04***		***p=***	***0.96***	***n=***	***3***	
IN, MW	**n!**	**r!**	**(n-r)!**	**p^r**	**qz^(n-r)**	**Individual Prob.**	**Cumulative Prob.**	**OUT, MW**
69	6	6	1	0.884736	1	0.884736	1	0
46	6	2	1	0.9216	0.04	0.110592	0.115264	23
23	6	1	2	0.96	0.0016	0.004608	0.004672	46
0	6	1	6	1	0.000064	0.000064	6.4E–05	69
						1		

System 2								
	FOR=	***0.04***		***p=***	***0.96***	***n=***	***2***	
IN, MW	**n!**	**r!**	**(n-r)!**	**p^r**	**q^(n-r)**	**Individual Prob.**	**Cumulative Prob.**	**OUT, MW**
24	2	2	1	0.9216	1	0.9216	1	0
12	2	1	1	0.96	0.04	0.0768	0.0784	12
0	2	1	2	1	0.0016	0.0016	0.0016	24

IN, MW	**Combine**	**Ind. Prob.**	**State**
93	69 + 24	0.815373	1
81	69 + 12	0.067948	2
69	69 + 0	0.001416	4
70	46 + 24	0.101922	3
58	46 + 12	0.008493	5
46	46 + 0	0.000177	7

IN, MW	Combine	Ind. Prob.	State
47	23 + 24	0.004247	6
35	23 + 12	0.000354	8
23	23 + 0	7.37E–06	10
24	0 + 24	5.9E–05	9
12	0 + 12	4.92E–06	11
0	0 + 0	1.02E–07	12

State	IN, MW	OUT, MW	Ind. Prob.
1	93	0	0.815373
2	81	12	0.067948
3	70	23	0.101922
4	69	24	0.001416
5	58	35	0.008493
6	47	46	0.004247
7	46	47	0.000177
8	35	58	0.000354
9	24	69	5.9E–05
10	23	70	7.37E–06
11	12	81	4.92E–06
12	0	93	1.02E–07

TABLE 5.5 Results, Example 5.5

OUT, MW	It's Prob.	After the State Before↓	After the State Before↓	Before the Next State↑	Before the Next State↑	Ind. Prob.
0	0.815372698	0	0	0	0	0.815372698
10	0	0	0	0.05435818	0	0.05435818
20	0	0.013589545	0	0.071345111	0.000849347	0.085784003
30	0	0.030576476	0.000566231	0.004246733	0	0.03538944
40	0	0.004246733	0	0.001698693	5.30842E–05	0.00599851
50	0	0.00254804	0.000123863	7.07789E–05	0	0.002742682
60	0	0.000283116	0	5.89824E–06	0	0.000289014
70	7.3728E–06	5.30842E–05	0	0	0	6.0457E–05
80	0	0	0	4.42368E–06	0	4.42368E–06
90	0	4.9152E–07	0	7.168E–08	0	5.632E–07
100	0	3.072E–08	0	0	0	3.072E–08
						1

Example (5.6):

Consider two systems 3 × 100 MW and 1 × 50 MW unit. Assuming that each unit having a FOR (U) = 0.05. Find the cumulative probabilities for each system. Then combine both systems together as one system and calculate the cumulative probabilities. After that, assuming the 100 MW-unit (the second system) is to be removed from the service using the deletion technique.

Solution: Tables 5.6a and 5.6b.

TABLE 5.6a Outage Probabilities Combining Both Systems

	System 1							
	FOR = 0.05			***p = 0.95***		***n = 3***		
IN, MW	**n!**	**r!**	**(n-r)!**	**p^r**	**q^(n-r)**	**Individual Prob.**	**Cumulative Prob.**	**OUT, MW**
300	6	6	1	0.857375	1	0.857375	1	0
200	6	2	1	0.9025	0.05	0.135375	0.142625	100
100	6	1	2	0.95	0.0025	0.007125	0.00725	200
0	6	1	6	1	0.00013	0.000125	0.000125	300
						1		
System 2								
	FOR = 0.05			***p = 0.95***		***n = 1***		
IN, MW	**n!**	**r!**	**(n-r)!**	**p^r**	**q^(n-r)**	**Individual Prob.**	**Cumulative Prob.**	**OUT, MW**
50	1	1	1	0.95	1	0.95	1	0
0	1	1	1	1	0.05	0.05	0.05	50

TABLE 5.6b Outage Probabilities Combining Both Systems

MW	CAP. IN (MW)	CAP. OUT (MW)	Individual Prob.	Cumulative Prob.
300 + 50	350	0	0.81450625	1
300 + 0	300	50	0.04286875	0.18549375
200 + 50	250	100	0.12860625	0.142625
200 + 0	200	150	0.00676875	0.01401875
100 + 50	150	200	0.00676875	0.00725
100 + 0	100	250	0.00035625	0.00048125
0 + 50	50	300	0.00011875	0.000125
0 + 0	0	350	0.00000625	0.00000625
			$\Sigma = 1$	

Any electric system has a number of generating-units. These units are scheduled periodically for maintenance or overhaul. In this case, the deletion technique is applied, as shown in the present example. When applying the deletion technique to remove the 50MW-unit from the capacity outage probability Table 5.6a is used by applying the following formula:

$$P'(X) = \frac{P(X) - P'(X-C).U}{1-U}$$

where X is the capacity, where $P'(X)$ and $P(X)$ are the cumulative probabilities of the capacity outage state of (X in MW) before and after adding the generating-unit.

$$\text{U} = \text{FOR} = 0.05$$

$$\text{C} = \text{Capacity Outage} = 50\text{MW}$$

$$P'(0) = \frac{1-(1)(0.05)}{0.95} = 1$$

$$P'(50) = \frac{0.18549375-(1)(0.05)}{0.95} = 0.142625$$

$$P'(100) = \frac{0.01401875-(0.142625)(0.05)}{0.95} = 0.00725$$

$$P'(150) = \frac{0.00048125-(0.00725)(0.05)}{0.95} = 0.000125$$

$$\text{Checking } P'(350) = \frac{0.00000625-(0.000125)(0.05)}{0.95} = 0.0$$

As conclusion: after deleting the 50MW-unit, the cumulative probabilities obtained are exactly what in the first considered system.

5.2.2 REPAIRABLE SYSTEM

Any repairable system has at least two-states, as shown in Figure 5.12. It might be represented by three-states, as shown in Figures 5.13a and 5.13b (Feng et al., 2018; Levitin et al., 2017; Li et al., 2018a).

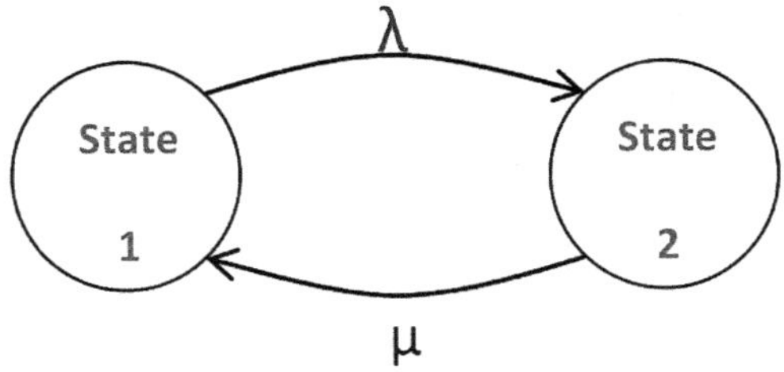

FIGURE 5.12 Two-state model.

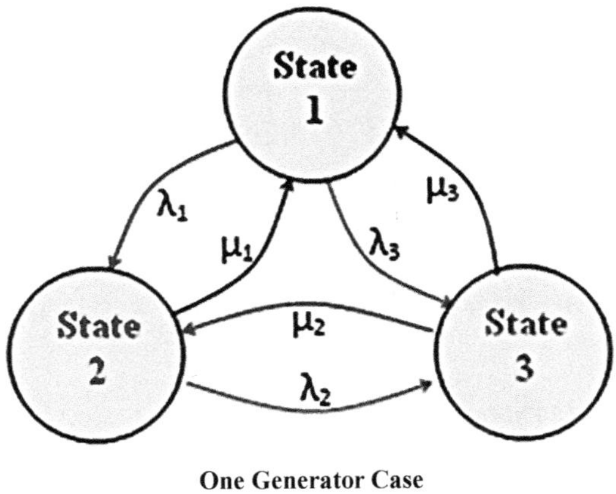

FIGURE 5.13a Three-state model.

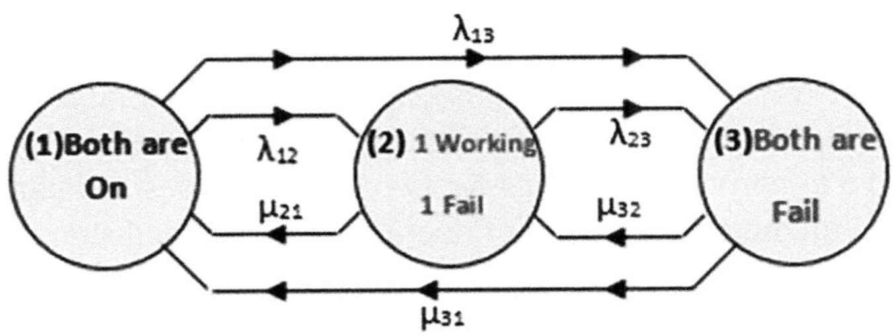

FIGURE 5.13b Three-state model.

5.2.3 UN-REPAIRABLE SYSTEM

Some systems might be considered as an un-repairable system. This system might be represented as illustrated in Figures 5.14a and 5.14b (Feng et al., 2018; Li et al., 2018b; Qingqing et al., 2018).

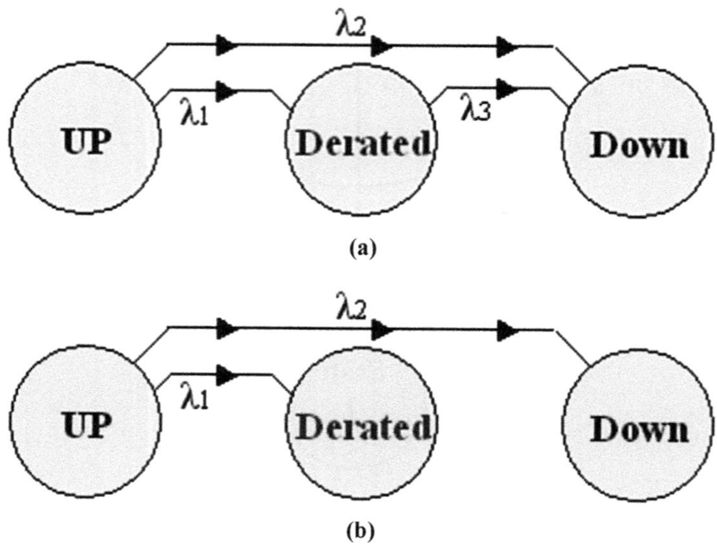

FIGURE 5.14 Three-state model (un-repairable).

5.2.4 RELIABILITY BLOCK DIAGRAM (RBD)

A reliability block diagram (RBD) performs any system that needs a reliability and availability analyses on large and complex systems (Billinton and Allan, 1992; Kim, 2011).

The presentation of the system will be a block diagram (Subburaj, 2015). This means to show network relationships. At the same time, the structure of the RBD defines the logical interaction of failures within a system that is required to sustain system operation. The Markov process is adaptive to derive expressions for two units in series and parallel. The two-state unit model conditions of fluctuating normal and stormy weather. The model is presented in a block-diagram. This model is used to simplify a system by the series and parallel block reduction. The RBD might be represented by including switch within the RBD. The RBD system is connected by a series or parallel configuration. The RBD can include components connected either in series or in parallel. It is well known that a failure in any series component causes the fail of the system. In a parallel-connected system, any component failure will not cause the failure of the system. The fail of the system needs all parallel components to fail. The availability is expressed as:

$$\text{Availability} = \frac{\text{System Up} - \text{Time}}{\text{System Up} - \text{Time} + \text{System Mean Down} - \text{Time}}$$

$$\therefore \text{ Availability} = \frac{\text{MTBF}}{\text{MTBF} + \text{MTTR}}$$

5.2.5 MARKOV MODELS

Markov models are a mathematical way of calculating steady-state and transient probabilities [x]. With the same pattern, the event frequencies in stochastic models are calculated. In Markov-model calculations, all life-times and repair times are assumed exponentially distributed. A Markov model consists of a number of states, with transition rates between the states that the model is passing through (Almuhaini and Al-Sakkaf, 2017; Billinton and Allan, 1992; IEEE Standard 493–2007).

5.3 STEADY-STATE AND TRANSIENT PROBABILITIES SOLUTION

Any model and if Markov model is considered, where it shows the transition rates between the states that the system is passing through shows the status of the model. Therefore, the probabilities the system is passing through will be shown and calculated through the times and known as transient probabilities. The final probabilities are known as the steady-state probabilities, where the system is reaching and will not be changed. These will be illustrated in the coming sub-sections (Billinton and Allan, 1992; El-Hay et al., 2019; Mahmood et al., 2018).

5.3.1 TWO-STATE MODEL

The two-state model represented by Figure 5.15 shows that State number 1 represents the Up-State, and State number 2 represents the Down-State. The transition rate from State (1) to State (2) is the failure-rate, where the transition rate from State (2) to State (1) is the repair-rate (Billinton and Allan, 1992; Groß et al., 2018).

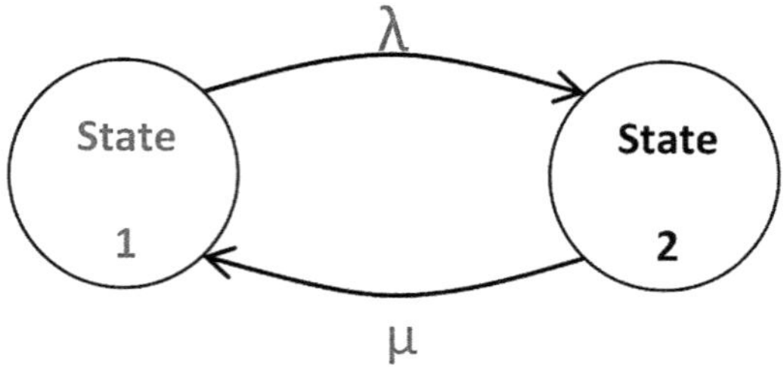

FIGURE 5.15 Two-states model.

$$P = \begin{bmatrix} 1-(\lambda-dt) & \lambda-dt \\ \mu-dt & 1-(\mu-dt) \end{bmatrix}$$

$$[P_1 \quad P_2] \begin{bmatrix} 1-\lambda dt & \lambda dt \\ \mu dt & 1-\mu dt \end{bmatrix} = [P_1 \quad P_2]$$

$$\begin{bmatrix} 1-\lambda dt & \lambda dt \\ \mu dt & 1-\mu dt \end{bmatrix} \begin{bmatrix} P_1 \\ P_2 \end{bmatrix} = [P_1 P_2]$$

$$(1 - \lambda dt)P_1 + \mu dt\, P_2 = P_1$$

$$\cancel{P_1} - \lambda dt\, P_1 + \mu dt\, P_2 = \cancel{P_1}$$

$$-\lambda dt\, P1 + \mu dt\, P2 = 0 \tag{1}$$

$$\lambda dt\, P_1 + (1-\mu dt)\, P_2 = P_2$$

$$\lambda dt\, P_1 + \cancel{P_2} - \mu dt\, P_2 = \cancel{P_2}$$

$$\lambda dt\, P_1 - \mu dt\, P_2 = 0 \tag{2}$$

Since the equations (1) and (2) are identical, one of them has to be replaced by another equation;

$$P_1(t) + P_2(t) = 1 \tag{3}$$

$$-(\lambda dt)\, P_1 + (\mu dt)\, P_2 = 0$$

$$-\lambda\, P_1 + \mu\, P_2 = 0 \tag{4}$$

Substituting (3) in (4)

$$-\lambda[1 - P_2] + \mu\, P_2 = 0$$

$$-\lambda + \lambda\, P_2 + \mu\, P_2 = 0$$

$$(\mu + \lambda)\, P2 = \lambda$$

$$\left.\begin{aligned} P_1 &= \mu / (\mu+\lambda) \\ P_2 &= \lambda / (\mu+\lambda) \end{aligned}\right\} \textit{Steady State Probabilities}$$

Transient Solution:

$$[P_1(t) \quad P_2(t)] \begin{bmatrix} (1-\lambda dt) & (\lambda dt) \\ (\mu dt) & (1-\mu dt) \end{bmatrix} = [P_1(t+dt) \qquad P_2(t+dt)]$$

$$(1 - \lambda dt)\, P_1(t) + (\mu dt)\, P_2(t) = P_1(t + dt) \tag{5}$$

$$(\lambda dt)\, P_1(t) + (1 - \mu dt)\, P_2(t) = P_2(t + dt) \tag{6}$$

$$P_1(t) - (\lambda dt)\, P_1(t) + (\mu dt)\, P_2(t) = P_1(t + dt) \tag{7}$$

$$(\lambda dt)\, P_1(t) + P_2(t) - (\mu dt)\, P_2(t) = P_2(t + dt) \tag{8}$$

$$-(\lambda dt)\, P_1(t) + (\mu dt)\, P_2(t) = \underbrace{P_1(t + dt) - P_1(t)}_{dP_1(t)} \tag{9}$$

$$(\lambda dt)\, P_1(t) - (\mu dt)\, P_2(t) = \underbrace{P_2(t + dt) - P_2(t)}_{dP_2(t)} \tag{10}$$

$$dP_1(t)/dt = -\lambda\, P_1(t) + \mu\, P_2(t) \tag{11}$$

$$dP_2(t)/dt = \lambda\, P_1(t) - \mu\, P_2(t) \tag{12}$$

Using Laplace transforms table (Appendix A) to convert from time-domain to S-domain and assuming that the initial conditions the generator (or system) is passing through are $P_1(0) = 1$ and $P_2(0) = 0$. This means that the first probability (i.e., the generator is perfect, where it has 100% probability), where the second probability (the generator failed, where it has 0% probability):

$$s.\, P_1(s) - P_1(0) = -\lambda\, P_1(s) + \mu\, P_2(s) \tag{13}$$

$$s.\, P_1(s) - 1 = -\lambda\, P_1(s) + \mu\, P_2(s)$$

$$s.\, P_2(s) - P_2(0) = \lambda\, P_1(s) - \mu\, P_2(s) \tag{14}$$

$$s.\, P_2(s) - 0 = \lambda\, P_1(s) - \mu\, P_2(s)$$

$$s.\, P_1(s) - 1 = -\lambda\, P_1(s) + \mu\, P_2(s) \tag{15}$$

$$s.\, P_2(s) = \lambda\, P_1(s) - \mu\, P_2(s) \tag{16}$$

$$[s + \lambda]\, P_1(s) - 1 = \mu\, P_2(s) \tag{17}$$

$$[s + \mu]\, P_2(s) = \lambda\, P_1(s) \tag{18}$$

From equation (14):

$$P_1(s) = [(s + \mu)/ \lambda]. P_2(s) \tag{19}$$

Substituting (19) in (17):

$$[(s + \lambda) (s + \mu) / \lambda]. P_2(s) = 1 + \mu P_2(s)$$

$$P_2(s) = \lambda / [(s + \lambda) (s + \mu) - \mu \lambda]$$

$$P_2(s) = \lambda / [s^2 + (\lambda+\mu)s + \cancel{\mu\lambda} - \cancel{\mu\lambda}]$$

$$P_2(s) = \lambda / [s(s+\mu+\lambda)]$$

$$P_2(s) = A/s + B/(s+\mu+\lambda)$$

$$A= \lambda / (\mu+\lambda)$$

$$B= -\lambda / (\mu+\lambda)$$

$$P_2(s) = [\lambda/(\mu+\lambda)]/s - [\lambda/(\mu+\lambda)]/[s+(\mu+\lambda)]$$

$$P_2(t) = [\lambda/(\mu+\lambda)] - [\lambda/(\mu+\lambda)]e^{-(\mu+\lambda)t}$$

$$P_1(t) = 1 - P_2(t)$$

In case the generator is becoming old, the first probability (i.e., the generator is becoming less functioning, where it has less than 100% probability). Assuming that it is becoming a 95% probability. In this case, the second probability becomes a 5% probability (the generator has a probability of 0.05). Therefore, equations (11) and (12) after applying Laplace transforms and by converting them from time-domain to S-domain becoming:

$$s. P_1(s)–P_1(0) = – \lambda P_1(s) + \mu P_2(s) \tag{20}$$

$$s. P_1(s)–0.95 = – \lambda P_1(s) + \mu P_2(s)$$

$$s. P_2(s)–P_2(0) = \lambda P_1(s) – \mu P_2(s) \tag{21}$$

$$s. P_2(s)–0.05= \lambda P_1(s) – \mu P_2(s)$$

$$s. P_1(s)–0.95 = – \lambda P_1(s) + \mu P_2(s) \tag{22}$$

$$s. P_2(s)–0.05= \lambda P_1(s) – \mu P_2(s) \tag{23}$$

$$[s + \lambda] P_1(s)–0.95 = \mu P_2(s) \tag{24}$$

$$[s + \mu] P_2(s)–0.05= \lambda P_1(s) \tag{25}$$

$$P_1(s)=\frac{1}{\lambda}\{(s+\mu)P_2(s)-0.05\} \tag{26}$$

Substituting (26) in (24):

$$(1/\lambda)[P_2(s)\,(s+\mu)-0.05][s+\lambda]=0.95+\mu\,P_2(s)$$

$$[(s+\lambda)\,(s+\mu)\,/\,\lambda]\,P_2(s)-(0.05\,[s+\lambda]/\lambda)=0.95+\mu\,P_2(s)$$

$$[\frac{(s+\lambda)\,(s+\mu)}{\lambda}P_2(s)-[\frac{0.05\,[s+\lambda]}{\lambda}=0.95+\mu\,P_2(s)$$

$$\left\{\frac{(s+\lambda)(s+\mu)-\mu\lambda}{\lambda}\right\}P_2(s)=\frac{0.05\,[s+\lambda]+0.95\lambda}{\lambda}$$

$$P_2(s)=\frac{0.05(s)+\lambda}{s^2+(\mu+\lambda)s}$$

$$P_2(s)=\frac{0.05(s)+\lambda}{s[s+(\mu+\lambda)]}$$

$$P_2(s)=\frac{0.05}{s+(\mu+\lambda)}+\frac{\lambda}{s[s+(\mu+\lambda)]}$$

The second term of the equation is $\frac{\lambda}{s[s+(\mu+\lambda)]}$

This term needs to be solved as:

$$\frac{\lambda}{s[s+(\mu+\lambda)]}=\frac{A}{s}+\frac{B}{s[s+(\mu+\lambda)]}$$

Therefore,

$$A\ (\text{when } s=0)=\frac{\lambda}{0+(\mu+\lambda)}$$

$$=\frac{\lambda}{(\mu+\lambda)}$$

and B [when s = – (μ + λ)] = $\frac{\lambda}{s}$

$$B=\frac{\lambda}{-(\mu+\lambda)}=\frac{-\lambda}{(\mu+\lambda)}$$

$$P_2(s) = \frac{0.05}{s+(\mu+\lambda)} + \frac{\lambda}{(\mu+\lambda)]}\Big]\frac{1}{s} - \left[\frac{\lambda}{\mu+\lambda}\right]\frac{1}{[s+(\mu+\lambda)]}$$

Then, converting the equation of $P_2(s)$ from the s-domain to t-domain:

$$P_2(t) = 0.05\, e^{-(\mu+\lambda)t} + \frac{\lambda}{(\mu+\lambda)} - \frac{\lambda}{(\mu+\lambda)} e^{-(\mu+\lambda)t}$$

$$P_2(t) = \frac{\lambda}{(\mu+\lambda)} + 0.05 - \frac{\lambda}{(\mu+\lambda)} e^{-(\mu+\lambda)t}$$

Substituting $P_2(s)$ in equation (26) to find $P_1(s)$:

$$P_1(s) = \frac{1}{\lambda}\{(s+\mu)\frac{1}{\mu}[(s+\lambda)P_1(s) - 0.95]\}$$

$$P_1(s) = \frac{(s+\mu)}{\mu\lambda}\{[(s+\lambda)P_1(s) - 0.95]\}$$

$$P_1(s)\frac{\mu\lambda}{(s+\mu)} = (s+\lambda)P_1(s) - 0.95$$

$$P_1(s)(s+\lambda) - \frac{\mu\lambda}{(s+\mu)}\} = 0.95$$

$$P_1(s)\{\frac{(s+\lambda)(s+\mu) - \mu\lambda}{(s+\mu)}\} = 0.95$$

$$P_1(s)\{\frac{s^2 + (\lambda s) + (\mu s) + \mu\lambda - \mu\lambda}{(s+\mu)}\} = 0.95$$

$$P_1(s)\frac{s[s+(\lambda)+(\mu)]}{(s+\mu)}\} = 0.95$$

$$P_1(s) = \frac{0.95(s+\mu)}{s[s+\mu+\lambda]}$$

$$P_1(s) = \frac{0.95\, s}{s[s+\mu+\lambda]} + \frac{0.95(\mu)}{s[s+\mu+\lambda]}$$

$$P_1(s) = \frac{0.95}{[s+\mu+\lambda]} + \frac{0.95(\mu)}{s[s+\mu+\lambda]}$$

Converting $\frac{0.95}{[s+\mu+\lambda]}$ from the s-domain to t-domain. It becomes 0.95 $e^{-(\mu+\lambda)t}$. The second term of the $P_1(s)$ equation can be converted and solved as:

$$\frac{0.95(\mu)}{s[s+(\mu+\lambda)]}=\frac{A}{s}+\frac{B}{[s+(\mu+\lambda)]}$$

Therefore, A (when s = 0) =

$$=\frac{0.95\ \mu}{(\mu+\lambda)}$$

and B [when s = –(μ + λ)] = $\frac{0.95\mu}{s}$

$$B=\frac{0.95\mu}{-(\mu+\lambda)}=\frac{-0.95\mu}{(\mu+\lambda)}$$

Substituting A and B in the second term. Then,

$$\frac{0.95(\mu)}{s[s+(\mu+\lambda)]}=\frac{0.95\ \mu}{(\mu+\lambda)}-\{\frac{0.95\mu}{(\mu+\lambda)}.\frac{1}{[s+(\mu+\lambda)]}\}$$

Now, converting the results from the s-domain to the t-domain. It becomes:

$$\frac{0.95\ \mu}{(\mu+\lambda)}-\{\frac{0.95\mu}{(\mu+\lambda)}e^{-(\mu+\lambda)\,t}\}$$

Therefore,

$$P_1(t)=0.95\ e^{-(\mu+\lambda)t}+\frac{0.95\mu}{(\mu+\lambda)}-\frac{0.95\mu}{(\mu+\lambda)}]\,e^{-(\mu+\lambda)t}$$

$$P_1(t)=\frac{0.95\mu}{(\mu+\lambda)}+0.95\ e^{-(\mu+\lambda)t}\,[1-\frac{\mu}{(\mu+\lambda)}]$$

5.3.2 THREE-STATE MODEL

Assuming that two identical generators considered, as shown in Figure 5.16 (Billinton and Allan, 1992; Groß et al., 2018).

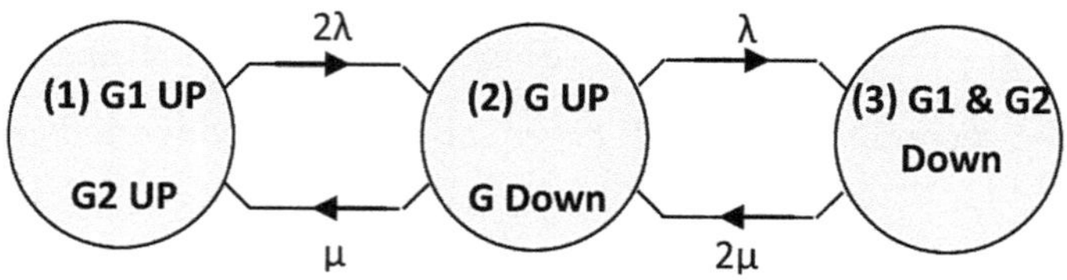

FIGURE 5.16 Three-states model as two identical generators case.

Where each failure rate is defined as (λ), and their repairing rates are (μ). The three states are defined as:

S1: Both generators are working.
S2: One generator is working, and the other is failed.
S2 and S1: Both generators are failed.

1) Find the steady-state solution.
2) Find the transient solution using the Laplace transforms.

where:

$$P_1(0) = 1$$
$$P_2(0) = 0$$
$$P_3(0) = 0$$

To find the steady-state probabilities, the following steps are followed:

$$P = \begin{bmatrix} 1-2\,\lambda.dt & \mu.dt & 0 \\ 2\,\lambda.dt & 1-(\mu+\lambda).dt & 2\,\mu.dt \\ 0 & \lambda.dt & 1-2\,\mu.dt \end{bmatrix}$$

$$P = \begin{bmatrix} 1-2\,\lambda.dt & \mu.dt & 0 \\ 2\,\lambda.dt & 1-(\mu+\lambda).dt & 2\,\mu.dt \\ 0 & \lambda.dt & 1-2\,\mu.dt \end{bmatrix} \begin{bmatrix} P_1 \\ P_2 \\ P_3 \end{bmatrix} = \begin{bmatrix} P_1 \\ P_2 \\ P_3 \end{bmatrix}$$

Let dt = 1

$$(1 - 2\,\lambda)\,P_1 + (\mu)\,P_2 = P_1$$

$$-2\,\lambda\,P_1 + \mu\,P_2 = 0 \qquad (27)$$

$(2\,\lambda)\, P_1 + [1 - (\mu + \lambda)]\, P_2 + [2\mu]\, P_3 = P_2$

$$2\,\lambda\, P_1 - (\mu + \lambda)\, P_2 + 2\mu\, P_3 = 0 \quad (28)$$

$[\lambda]\, P_2 + [1 - 2\mu]\, P_3 = P_3$

$$\lambda\, P_2 - 2\mu\, P_3 = 0 \quad (29)$$

From Equations (27) and (29)

$$P_1 = (\mu/2\lambda)\, P_2 \qquad \text{and} \qquad P_3 = (\lambda/2\mu)\, P_2 \quad (30)$$

Equation: $P_1 + P_2 + P_3 = 1$ replacing equation (28)

$$(\mu/2\lambda)\, P_2 + P_2 + (\lambda/2\mu)\, P_2 = 1$$

$$P_2\, [(\mu/2\lambda) + 1 + (\lambda/2\mu)] = 1$$

Therefore,

$$P_2 = 1/\, [(\mu^2 + 2\lambda\mu + \lambda)/2\lambda\mu] = 2\lambda\mu \,/\, (\mu + \lambda)^2$$

$$P_1 = \mu \,/\, (2\lambda).\, (2\mu\lambda) \,/\, (\mu + \lambda)^2 = \mu^2/(\mu + \lambda)^2$$

$$P_3 = (\lambda/2\mu)\, (2\lambda\mu/(\mu+\lambda)^2) = \lambda^2/(\mu+\lambda)^2$$

Transient probabilities:

$$\begin{bmatrix} 1-2\,\lambda.dt & \mu.dt & 0 \\ 2\,\lambda.dt & 1-(\mu+\lambda).dt & 2\,\mu.dt \\ 0 & \lambda.dt & 1-2\,\mu.dt \end{bmatrix} \begin{bmatrix} P_1(t) \\ P_2(t) \\ P_3(t) \end{bmatrix} = \begin{bmatrix} P_1(t+dt) \\ P_2(t+dt) \\ P_3(t+dt) \end{bmatrix}$$

$(1 - 2\,\lambda dt).\, P_1(t) + (\mu.dt).\, P_2(t) = P_1(t + dt)$

$$(-2\,\lambda.dt).\, P_1(t) + (\mu.dt).\, P_2(t) = P_1\underbrace{(t + dt) - P1(t)}_{dP_1(t)}$$

$$\frac{dP_1(t)}{dt} = -2\lambda\, P_1(t) + \mu\, P_2(t)$$

$(2\,\lambda.dt).\, P_1(t) + [1 - (\mu + \lambda)dt]\, P_2(t) + [2\mu\, dt]\, P_3(t) = P_2(t + dt)$

$$(2\,\lambda.dt).\, P_1(t) - [(\mu + \lambda)dt]\, P_2(t) + [2\mu\, dt]\, P_3(t) = P_2\underbrace{(t + dt) - P_2(t)}_{dP_2(t)}$$

$$\frac{dP_2(t)}{dt} = 2\lambda\ P_1(t) - (\mu + \lambda) P_2(t) + 2\mu\ P_3(t)$$

$[\lambda.dt]\ P_2(t) + [1 - 2\mu\ dt]\ P_3(t) = P_3(t + dt)$

$[\lambda.dt]\ P_2(t) - [2\mu\ dt]\ P_3(t) = P_3(t + dt) - \underbrace{P_3(t + dt) - P_3(t)}_{dP_3(t)}$

$$\frac{dP_3(t)}{dt} = \lambda\ P_2(t) - 2\mu P_3(t)$$

Using Laplace transform (Appendix A) with the following initial values:

$P_1(0) = 1$, $P_2(0) = 0$ and $P_3(0) = 0$

$S.P_1(s) - P_1(0) = -2\lambda.\ P_1(s) + \mu\ P_2(s)$

$S.P_1(s) - 1 = -2\lambda.\ P_1(s) + \mu\ P_2(s)$

$$P_1(s) = [\mu P_2(s) + 1]/(s + 2\lambda) \tag{31}$$

$S.P_2(s) - P_2(0) = 2\lambda\ P_1(s) - (\mu + \lambda).P_3(s) + 2\ \mu\ P_3(s)$

$$P_2(s) = [2\lambda\ P_1(s) + 2\mu\ P_3(s)]/[s + (\mu + \lambda)] \tag{32}$$

$S.P_3(s) - P_3(0) = \lambda.P_2 - 2\mu.P_3(s)$

$(s + 2\mu).P_3(s) = \lambda.P_2(s)$

$$P_3(s) = \lambda/[s + 2\mu].P_2(s) \tag{33}$$

Substitute equations (31) and (33) in (32):

$P_2(s) = [(2\lambda)\ [(\mu P_2(s) + 1)/(s + 2\ \lambda)] + [(2\mu\lambda P_2(s)/s + 2\mu)]]/(s + \mu + \lambda)$

$P_2(s) = [2\lambda\mu P_2(s) + 2\lambda]/(s + 2\lambda)(s + \mu + \lambda)] + [(2\mu\lambda P_2(s)/(s + 2\mu)(s + \mu + \lambda)$

$(s + \mu + \lambda)P_2(s) - [2\lambda\mu P_2(s)/(\cancel{s + 2\lambda})(s + \mu + \lambda)] - [(2\mu\lambda P_2(s)/(\cancel{s + 2\mu})$
$(s + \mu + \lambda) = \cancel{2\lambda/(s + 2}\ \lambda)(s + \mu + \lambda)$

$P_2(s)[s + \mu + \lambda - (2\lambda\mu)/(s + 2\lambda) - (2\lambda\mu)/(s + 2\mu)] = 2\lambda/(s + 2\lambda)$

$P_2(s)[(s + 2\lambda)P_2(s) + 2\mu)(s + \mu + \lambda) - 2\lambda(s + 2\mu) - 2\cancel{\mu\lambda(s + 2\lambda)}]/$
$[(s + 2\lambda)(s \cancel{+ 2\mu)]} = [2\lambda/(s + 2\lambda)]$

$P_2(s) = [2\lambda(s + 2\mu)]/[(s + 2\lambda)(s + 2\mu)(s + \mu + \lambda) - 2\mu\lambda s - 4\mu^2\lambda$
$- 2\mu\lambda s - 4\mu]$

$P_2(s) = [2\lambda(s + 2\mu)]/[-4\mu\lambda(s + \mu + \lambda) + (s^2 + 2s\mu + 2s\lambda + 4\mu\lambda)(s + \lambda + \mu)]$

$$P_2(s) = [2\lambda(s + 2\mu)]/[(s + \mu + \lambda)(-4\mu\lambda + 4\mu\lambda + s^2 + 2s\mu + 2s\lambda)]$$

$$P_2(s) = [2\lambda(s + 2\mu)]/[(s + \mu + \lambda)(s^2 + 2s(\mu + \lambda))]$$

$$P_2(s) = [2\lambda(s + 2\mu)]/[(s + \mu + \lambda)s(s + 2(\mu + \lambda))]$$

$$P_2(s) = A/(s + (\mu + \lambda)) + B/s + C/[s + 2(\mu + \lambda)]$$

$$2\lambda(s + \mu) = As(s + 2(\mu + \lambda)) + B(s + \mu + \lambda)(s + 2(\mu + \lambda) + Cs(s + \mu + \lambda)$$

Let $s=0 \Longrightarrow 4\,\mu\lambda = B(\lambda + \mu)(2(\lambda + \mu) \Longrightarrow B = 4\mu\lambda/(\lambda + \mu)^2$
$B = 2\mu\lambda/(\lambda + \mu)^2$

Let $s = -\mu - \lambda \Longrightarrow 2\lambda(-(\lambda + \mu)) + (2\mu) = -A(\lambda + \mu)(-\lambda + \mu) + 2(\lambda + \mu)$
$2\lambda(\mu - \lambda) = -A(\lambda + \mu)(\lambda + \mu)$
$A = 2\lambda(\mu - \lambda)/(\lambda + \mu)^2$

Let $s = -2(\lambda + \mu) \Longrightarrow 2\lambda(-2(\lambda + \mu) + 2\mu) = C(-2(\lambda + \mu)\,(-2(\lambda + \mu)$
$+ \mu + \lambda$
$2\lambda(-2\lambda - 2\mu + 2\mu) = -2C(\lambda + \mu)(-2\lambda - 2\mu + \mu + \lambda)$
$-4\lambda^2 = -2C(\lambda + \mu)(-(\mu + \lambda))$
$C = 2\lambda^2/(\lambda + \mu)^2$

$$P_2(s) = [-2\lambda(\mu - \lambda)/(\lambda + \mu)^2]/[s + (\mu + \lambda)] + [2\lambda\mu/(\lambda + \mu)^2/s] - [2\lambda^2/(\lambda + \mu)^2/(s + 2(\lambda + \mu)]$$

$$P_2(t) = \frac{2\mu\lambda}{(\lambda+\mu)^2} + \frac{(2\lambda^2 - 2\mu\lambda)}{(\lambda+\mu)^2} e^{-(\mu+\lambda)t} - 2\frac{\lambda^2}{(\lambda+\mu)^2} e^{-2(\mu+\lambda)t}$$

$$P_2(t) = \frac{2\mu\lambda}{(\lambda+\mu)^2} + \frac{2\lambda(\lambda-\mu)}{(\lambda+\mu)^2} e^{-(\mu+\lambda)t} - 2\frac{\lambda^2}{(\lambda+\mu)^2} e^{-2(\mu+\lambda)t}$$

$$P_3(s) = \lambda/(s + 2\mu)\,P2(s)$$

$$P_3(s) = \lambda/(s + 2\mu)[-2\lambda(\mu - \lambda)/(\lambda + \mu)^2]/[s + (\mu + \lambda)] + \lambda/(s + 2\mu)\,[2\lambda\mu/(\lambda + \mu)^2/s] - \lambda/(s + 2\mu)\,[2\lambda^2/(\lambda + \mu)^2/(s + 2(\lambda + \mu)]$$

$$P_3(s) = -2\lambda^2(\mu - \lambda)/(\lambda + \mu)^2]/[(s + 2\mu)(s + \mu + \lambda)] + [2\lambda\mu/((\lambda + \mu)^2/(s(s + 2\mu)] - 2\lambda^3/(\mu + \lambda)^2/(s + 2\mu)(s + 2(\lambda + \mu))]$$

$$P_3(s) = [A/\ s + 2\mu] + [B/(s + (\mu + \lambda))] + C/s + [D/(s + 2\mu)] - [E/(s + 2\mu)] + F/(s + 2(\lambda + \mu)]$$

$-2\lambda^2(\mu - \lambda)/(\lambda + \mu)^2/(s + 2\mu)(s + \lambda + \mu) = A/(s + 2\mu) + B/[s + (\mu + \lambda)]$

$-2\lambda^2(\mu - \lambda)/(\lambda + \mu)^2 = A(s + \mu + \lambda) + B(s + 2\mu)$

Let $s= -2\mu$

$-2\lambda^2(\mu - \lambda)/(\lambda + \mu)^2 = A(-2\mu + \mu + \lambda) + B(-2\mu + 2\mu)$

$-2\lambda^2(\mu - \lambda)/(\lambda + \mu)^2 = -A(\mu - \lambda)$

$A = 2\lambda^2/(\lambda + \mu)^2$

Let $s= -(\mu + \lambda)$

$-2\lambda^2(\mu - \lambda)/(\lambda + \mu)^2 = B(-\mu - \lambda + 2\mu)$

$-2\lambda^2(\mu - \lambda)/(\lambda + \mu)^2 = B(-\mu - \lambda)$

$B = -2\lambda^2/(\lambda + \mu)^2$

$2\mu\lambda^2/(\lambda + \mu)^2/s(s + 2\mu) = C/s + D(s + 2\mu)$

$2\mu\lambda^2/(\lambda + \mu)^2 = C(s + 2\mu) + Ds$

Let $s= 0$

$2\mu\lambda^2/(\lambda + \mu)^2 = C(2\mu)$

$C = \lambda^2/(\lambda + \mu)^2$

Let $s= -2\mu$

$2\mu\lambda^2/(\lambda + \mu)^2 = -2\mu D$

$D = -\lambda^2/(\lambda + \mu)^2$

$-2\lambda^2/(\lambda + \mu)^2/[(s + 2\mu)(s + 2(\lambda + \mu))] = E/(s + 2\mu) + F/(s + 2(\lambda + \mu))$

$-2\lambda^2/(\lambda + \mu)^2 = E(s + 2(\lambda + \mu)) + F(s + 2\mu)$

Let $s = -2\mu$

$-2\lambda^2/(\lambda + \mu)^2 = E(-2\mu + 2\lambda + 2\mu)$

$E = -\lambda^2/(\lambda + \mu)^2$

Let $s = -2(\lambda + \mu)$

$-2\lambda^2/(\lambda + \mu)^2 = F(-2\lambda - 2\mu + 2\mu)$

$F = \lambda^2/(\lambda + \mu)^2$

$P_3(s) = 2\lambda^2/(\lambda + \mu)^2/(s + 2\mu) + (-2\lambda^2/(\lambda + \mu)^2/(s + (\mu + \lambda)) + \lambda^2/(\lambda + \mu)^2/s +$
$(-\lambda^2/(\lambda + \mu)^2/(s + 2\mu) - (-\lambda^2/(\lambda + \mu)^2/(s + 2\mu) - (\lambda^2/(\lambda + \mu)^2/(s + 2(\mu + \lambda))$

$P_3(s) = 2\lambda^2/(\lambda + \mu)^2/(s + 2\mu) - (2\lambda^2/(\lambda + \mu)^2/(s + (\mu + \lambda)) + \lambda^2/(\lambda + \mu)^2/s -$
$(\lambda^2/(\lambda + \mu)^2/(s + 2\mu + \lambda)$

$$P_3(t) = \frac{2\lambda^2}{(\lambda+\mu)^2} e^{-2(\mu+\lambda)t} - \frac{2\lambda^2}{(\lambda+\mu)^2} e^{-(\mu+\lambda)t} - \frac{\lambda^2}{(\lambda+\mu)^2} e^{-2(\mu+\lambda)t} + \frac{\lambda^2}{(\lambda+\mu)^2}$$

$$P_3(t) = \frac{\lambda^2}{(\lambda+\mu)^2} e^{-2(\mu+\lambda)t} - \frac{2\lambda^2}{(\lambda+\mu)^2} e^{-(\mu+\lambda)t} + \frac{\lambda^2}{(\lambda+\mu)^2}$$

$P_1(t) = 1 - P_2(t) – P_3(t)$

$$P_1(t) = \frac{\lambda^2}{(\lambda+\mu)^2} e^{-2(\mu+\lambda)t} + \frac{2\lambda\mu}{(\lambda+\mu)^2} e^{-(\mu+\lambda)t} + \frac{\mu^2}{(\lambda+\mu)^2}$$

5.4 SERIES COMPONENTS

For any series-system (Zhao et al., 2017) as having two components A and B, as shown in Figure 5.17:

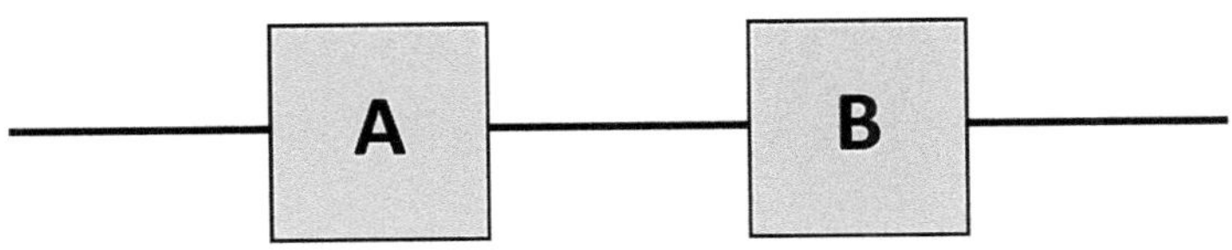

FIGURE 5.17 Two components in series.

If component A has a reliability R_A and the second component B has a reliability R_B, therefore, the overall reliability is:

$$R_{overall} = R_{total} = R_A . R_B$$

$$R_{overall}(t) = R_A(t). R_B(t)$$

Considering each component has a different failure rate, i.e., λ_A we and λ_B, therefore the total reliability becomes:

$R_{total}(t) = [e^{-\lambda At}] [e^{-\lambda Bt}]$

$$R_{total}(t) = e^{-(\lambda A + \lambda B)t}$$

Example (5.7):

Consider a system having 20 numbers of components in series. These components are identical, where the failure rate is 0.15 per year for each component.

Calculate the overall reliability for 2 years, 5 years, and 10,760 hours.

Solution:

The results obtained by calculation and plotted as shown in Figure 5.18.

$$R_{total}\ (t = 2years) = e^{-(20 \times 0.15)2} = 0.002478$$

$$R_{total}\ (t = 5years) = e^{-(20 \times 0.15)5} = 0.000000306$$

$$R_{total}\ (t = 1.23years) = e^{-(20 \times 0.15)1.23} = 0.025$$

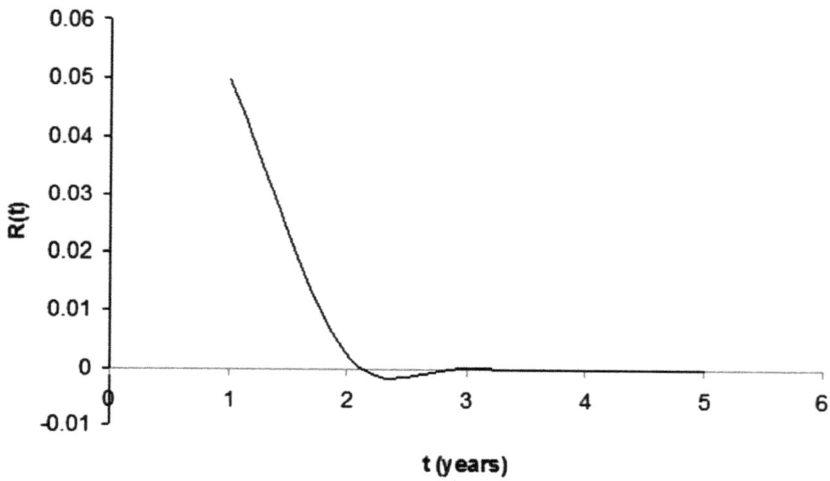

FIGURE 5.18 Reliabilities for the example (5.7).

For the same periods, find the unreliability (Table 5.7).

TABLE 5.7 Results for Example 5.7

Time (years)	Reliability R(t)	Un-reliability Q(t) = 1 – R(t)
1.23	0.0251	0.9749
2	0.002478	0.997522
5	0.000000306	0.999999694

Example (5.8):

A power system has 150 identical components. These components are connected in series. The system was running for 15 years, and its reliability is 0.99 at this time, which is very reliable. This means that the system is very efficient. Calculate the reliability of each component at that time.

Solution:

$$R_{Total}(t) = R_1(t)R_2(t)..............R_{150}(t)$$

Since the components are identical. Therefore,

$$R_{Total}(t) = e^{-150\lambda t}$$

$$0.99 = e^{-150\lambda t}$$

$$0.99 = e^{-(150 \times \lambda)(15)}$$

Taking the (ln) for both sides

$$\ln(0.99) = -(150 \text{x} \lambda)(15)$$

$\therefore \lambda = 4.4668159 \text{ x}10^{-6}$ per year

Then,

$$R_1(t) = e^{-\lambda 1 t}$$

$$R_1(t) = e^{-(4.4668159 \text{ x}10^{-6})t}$$

$$R_1(t) = 0.999933$$

Another Solution:

$$R_{Total}(t) = R_1(t)R_2(t)..............R_{150}(t)$$

$$R_{Total}(t) = R^{150}(t)$$

$$0.99 = R^{150}(t)$$

$$\ln(0.99) = (150)[\ln R(t)]$$

$$\ln[R(t)] = -(6.7002239 \text{x}10^{-5})$$

Taking the (exponential) for both sides:

$$R(t) = 0.999933$$

Since, the components are identical. Therefore, the reliability of each component is equal to 0.999933.

Example (5.9):

The electric power system has 170 identical components connected in series. The system was running for 17 years, and its overall reliability is 0.92 at this time, which is very reliable. This means that the system is very efficient. Calculate the reliability of each component at that time.

Solution:

$$R_{Total}(t) = R_1(t)R_2(t)..............R_{150}(t)$$

Since the components are identical. Therefore,

$$R_{Total}(t) = e^{-170\lambda t}$$

$$0.92 = e^{-(170\lambda)t}$$

$$0.92 = e^{-(170\times\lambda)(17)}$$

Taking the (ln) for both sides

$$\ln(0.92) = -(150\,\text{x}\lambda)(17)$$

$\therefore \lambda = 2.88518\ \text{x}10^{-5}$ per year

Then,

$$R_1(t) = e^{-\lambda 1 t}$$

$$R_1(t) = e^{-\left(2.88518\ \text{x}10^{-5}\right)t}$$

$$R_1(t) = 0.99950964$$

Another Solution:

$$R_{Total}(t) = R_1(t)R_2(t)..............R_{150}(t)$$

$$R_{Total}(t) = R^{170}(t)$$

$$0.92 = R^{170}(t)$$

$$\ln(0.92) = (170)[\ln R(t)]$$

$$R(t) = 0.99950964$$

Since, the components are identical. Therefore, the reliability of each component is equal to 0.99950964.

5.5 PARALLEL COMPONENTS

If we have two parallel components illustrated in Figure 5.19 (Ota and Kimura, 2017).

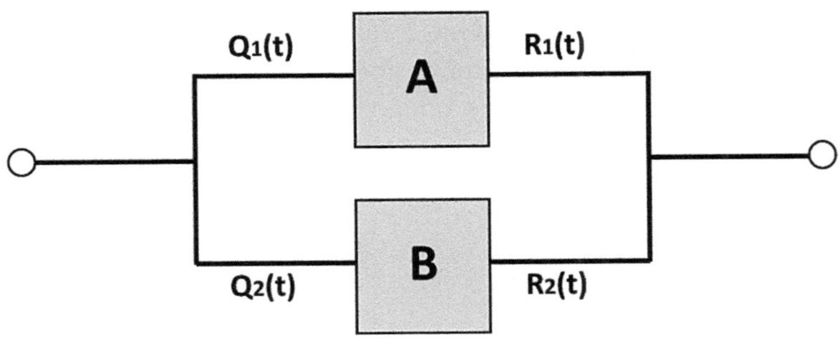

FIGURE 5.19 Two components in parallel.

$Q_T(t) = Q_1(t).\ Q_2(t)$

where: Q(t) is the un-reliability of the system.

$$\begin{aligned} R_T(t) &= 1 - Q_T(t) \\ &= 1 - [Q_1(t).\ Q_2(t)] \\ &= 1 - [(1 - R_1(t)).\ (1 - R_2(t))] \\ &= 1 - [1 - R_1(t) - R_2(t) + (R_1(t).\ R_2(t))] \\ &= 1 - 1 + R_1(t) + R_2(t) - [R_1(t).\ R_2(t)] \\ &= R_1(t) + R_2(t) - [R_1\ (t).\ R_2(t)] \\ &= e^{-\lambda 1t} + e^{-\lambda 2t} - [e^{-\lambda 1t}.\ e^{-\lambda 1t}] \\ &= e^{-\lambda 1t} + e^{-\lambda 2t} - e^{-(\lambda 1 + \lambda 2)t} \end{aligned}$$

When we have n-number of parallel systems, the un-reliability is:

$$Q_T(t) = Q_1(t).\ Q_2(t).\ Q_3(t) \ldots\ldots\ldots Q_n(t)$$

$$Q_T(t) = \prod_{i=1}^{n} Q_j(t)$$

The overall reliability will be:

$$R_T(t) = 1 - \prod_{i=1}^{n} Q_j(n)$$

Example (5.10):

Two systems connected in parallel. The first has a failure-rate equal to 0.001 per hour, when the second has a failure-rate equal to 0.05 per hour, these two systems have been running for 10 hours. Calculate the overall reliability. Then if both systems connected in series, what is the total reliability?

Solution:

System #1: $\lambda_1 = 0.001/\text{hr}$
System #2: $\lambda_2 = 0.05/\text{hr}$

$t = 10$ hrs

1st connection (parallel):

$$R_{T1}\ (t = 10) = e^{-\lambda 1t} + e^{-\lambda 2t} - e^{-(\lambda 1 + \lambda 2)t}$$

$$R_{T1}\ (t = 10) = e^{-(0.001)10} + e^{-(0.05)10} - e^{-(0.001 - 0.05)10}$$

$$R_{T1}\ (t = 10) = 0.996$$

2nd connection (series):

$$R_{T2}\ (t = 10) = R_A(t).\ R_B(t)$$

$$R_{T2}\ (t = 10) = e^{-(\lambda 1 + \lambda 2)t}$$

$$R_{T2}\ (t = 10) = e^{-(0.001 + 0.05)10}$$

$$R_{T2}\ (t = 10) = 0.60049$$

Example (5.11):

Consider the previous system (both A&B); two cases are considered (parallel and series).

Calculate the total reliability for 5, 10, 15, 20, and 25 hours running time.

Solution:

The results are shown in Table 5.8 and Figures 5.20 and 5.21.

TABLE 5.8 Results, Example 5.11

Time (Hour)	$R_A(t)$	$R_B(t)$	$R_{T1}(t)$ Parallel	$R_{T2}(t)$ Series
5	0.995012	0.778801	0.9989	0.7749
10	0.990050	0.606531	0.9961	0.6605
15	0.985112	0.472367	0.9921	0.4653
20	0.980199	0.367879	0.9875	0.3606
25	0.975310	0.286505	0.9824	0.2794

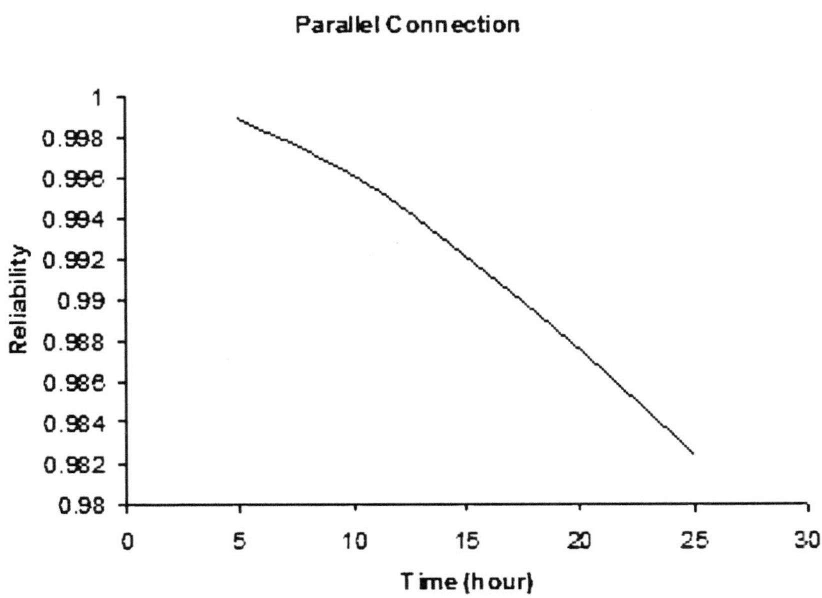

FIGURE 5.20 Results, example 5.11 (parallel connection case).

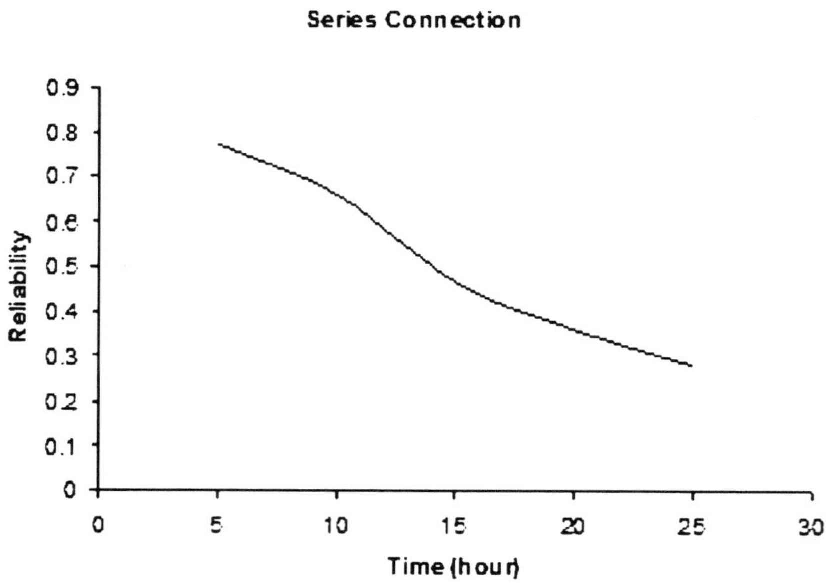

FIGURE 5.21 Results, example 5.11 (series connection case).

Example (5.12):

From your study of stochastic connections, consider two cases:

a. Two stochastic components connected in series with two outage rates λ_1 and λ_2, where the repair times are r_1 and r_2, respectively. Derive the expressions for outage rate λ_s and repair time r_s of the series connection in their final form.

b. The same components are reconnected in parallel, derive the expressions for outage rate λ_p and repair time r_p of the parallel connection.

Solution:

a. Two components connected in series

$$\lambda_s = \lambda_1 + \lambda_2$$

$$q_s = q_1 + q_2$$

$$= \lambda_1 r_1 + \lambda_2 r_2$$

Since, $q_s = \lambda_s r_s$

Therefore,

$$\lambda_s r_s = \lambda_1 r_1 + \lambda_2 r_2$$

Then,

$$r_s = \frac{\lambda 1 \, r1 + \lambda 2 \, r2}{\lambda 1 + \lambda 2}$$

$$\therefore \quad \lambda s = \sum_{i=1}^{n} \Sigma \lambda_i$$

and

$$r_s = \frac{\sum_{i=1}^{n} \lambda i \, ri}{\sum_{i=1}^{n} \lambda i}$$

b. Two components connected in series (the same components)

$$\lambda_p = \lambda_2 q_1 + \lambda_1 q_2$$

$$\lambda_p = \lambda_1 \lambda_2 r_1 + \lambda_1 \lambda_2 r_2$$

$$\lambda_p = \lambda_1 \lambda_2 (r_1 + r_2)$$

Since, the components are connected in parallel. Therefore,

$$q_p = q_{1.} \, q_2$$

$$= \lambda_1 r_1 \lambda_2 r_2$$

$$= \lambda_p r_p$$

$$\therefore r_p = \frac{\lambda_1 r_1 \lambda_2 r_2}{\lambda_1 \lambda_2 (r_1 + r_2)}$$

$$r_p = \frac{(r_1 r_2)}{(r_1 + r_2)}$$

Example (5.13):

Seven components are connected in series once then in parallel. These components are operated for 35 months, and their failure rates are as follows (Table 5.9):

TABLE 5.9 Example 5.13 Data

Component No.	Failure Rate (f/Month)
1	0.002
2	0.005
3	0.003
4	0.019
5	0.015
6	0.008
7	0.007

Calculate the overall reliability of the system for both cases.

Solution:

Reliability	Value
R_1	0.93239382
R_2	0.839457021
R_3	0.900324523
R_4	0.514273528
R_5	0.591555364
R_6	0.755783741
R_7	0.782704538

Un-reliability	Value
Q_1	0.06760618
Q_2	0.160542979
Q_3	0.099675477
Q_4	0.485726472
Q_5	0.408444636
Q_6	0.244216259
Q_7	0.217295462

The reliability and reliability calculations are illustrated below:

Series, Reliability (R)	0.126818291
Parallel, Un-reliability (Q)	1.13898E–05
Parallel, Reliability (R)	0.99998861

5.6 SERIES-PARALLEL COMPONENTS

In the real network, the connection of the components will take place in mixed connections. This means that the connection will be series in a branch and parallel in another branch, and so on. In the present section, a number of examples will be addressed in mixed connections (Feizabadi and Jahromi, 2018; Mo et al., 2018).

Example (5.14):

Find the overall reliability for the system between A and B, as shown in Figure 5.22. Then, calculate the total reliability at t = 5 hours.

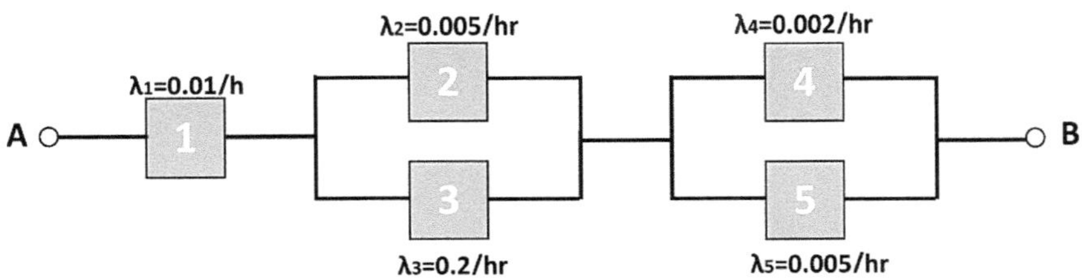

FIGURE 5.22 Series-parallel components.

Solution:

2//3 ⟹ 6

4//5 ⟹ 7

1, 6 and 7 are in series

$$R_1 = e^{-\lambda 1t} = e^{-0.01t}$$

$$R_6 = 2//3 = R_2 + R_3 - R_2.R_3$$
$$= e^{-0.005t} + e^{-0.2t} - e^{-0.205t}$$

$$R_7 = 4//5 = R_4 + R_5 - R_4.R_5$$
$$= e^{-0.002t} + e^{-0.005t} - e^{-0.007t}$$

$$R_{Total} = R_1.\ R_6.\ R_7$$
$$R_1\ (t = 5\ \text{hours}) = e^{-(0.01\times 5)} = 0.91229$$
$$R_6\ (t = 5\ \text{hours}) = e^{-(0.005\times 5)} + e^{-(0.2\times 5)} - e^{-(0.205\times 5)} = 0.98439$$

$$R_7\ (t = 5\ \text{hours}) = e^{-(0.002\times5)} + e^{-(0.005\times5)} - e^{-(0.007\times5)} = 0.99975$$

$$R_{Total} = R_1 .\ R_6 .\ R_7 = (0.91229).(0.98439).(0.99975) = 0.897828$$

Example (5.15):

Consider the system illustrated in Figure 5.23, and calculate the total reliability at 2, 5, and 7 years.

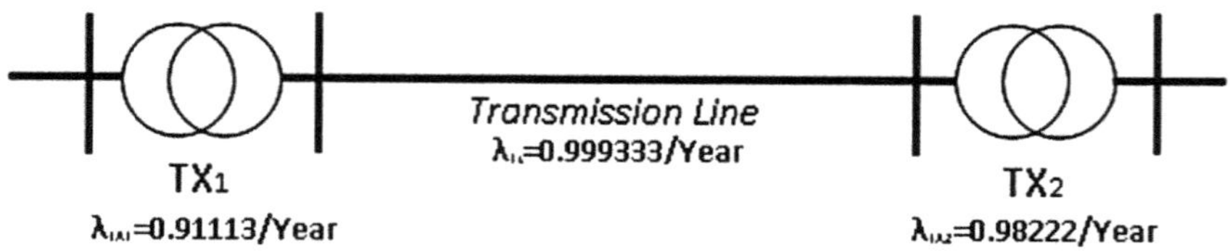

FIGURE 5.23 Simple electric power network.

Solution:

$$R_{Total}\ (t = 2\ \text{years}) = e^{-[(0.91113 + 0.999333 + 0.98222)\times2]} = 0.00307$$

$$R_{Total}\ (t = 5\ \text{years}) = e^{-[(0.91113 + 0.999333 + 0.98222)\times5]} = 0.52314 \times 10$$

$$R_{Total}\ (t = 7\ \text{years}) = e^{-[(0.91113 + 0.999333 + 0.98222)\times7]} = 1.607 \times 10$$

Example (5.16):

Consider a system having reliability equal to 0.89752 after six years of running. Find the reliability of the same system after 10 years.

Solution:

$$R_{Total}\ (t = 6\ \text{years}) = 0.89752$$

$$R_{Total}\ (t = 10\ \text{years}) = ?$$

$$R_{Total}\ (t = 6\ \text{years}) = e^{-\lambda t}$$

$$0.89752 = e^{-6\lambda}$$

$$\text{Ln}(0.89752) = -6\ \lambda$$

$$\lambda = 0.01802\ /\text{year}$$

$$R_{Total}\ (t = 10\ \text{years}) = e^{-(0.018026\times100)}$$

$$= 0.835103$$

Example (5.17):

A power system network shown in Figure 5.24 represents non-identical components. Derive the overall reliability formula between the points A and B for the network in terms of un-reliability for each component.

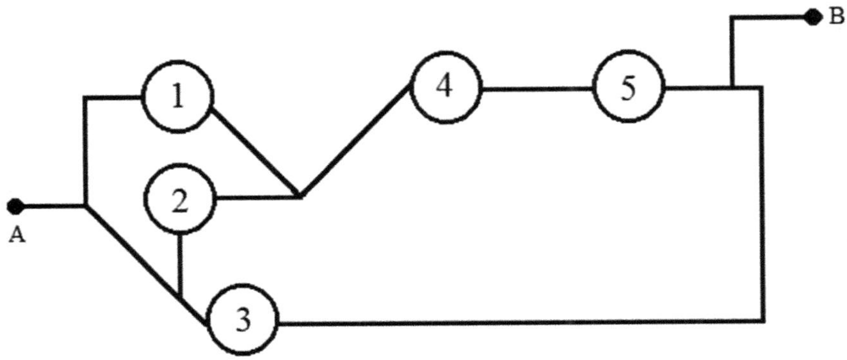

FIGURE 5.24 Non-identical components of a power system network.

Solution:

Arranging the system in a better form such as (Figure 5.25):

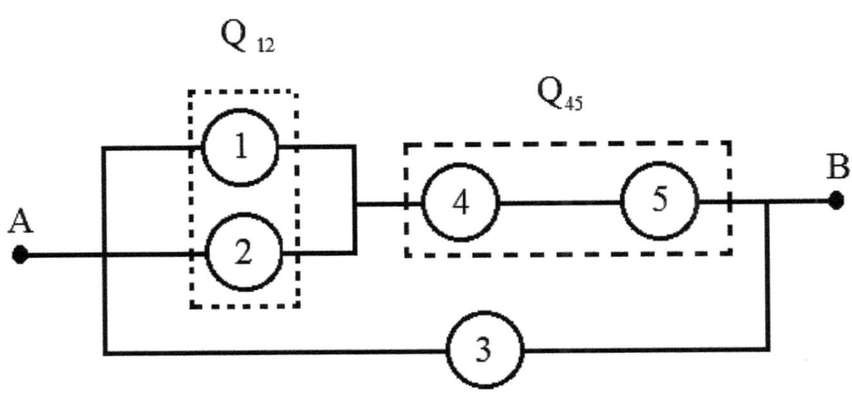

FIGURE 5.25 Arrangement of non-identical components of a power system network.

As stated, the components are non-identical. The components have un-reliabilities, respectively Q_1, Q_2, Q_3, Q_4, and Q_5.

Both components 1 and 2 are parallel. Therefore,

$$Q_{12}(t) = Q_1(t)\, Q_2(t)$$

Components 4 and 5 are series. Therefore,

$$R_{45}(t) = R_4(t)\, R_5(t)$$

$$\therefore Q_{45}(t) = 1 - R_{45}(t)$$

$$= 1 - [R_4(t)\, R_5(t)]$$

$$= 1 - [\{1 - Q_4(t)\}\{1 - Q_5(t)\}]$$

$$= 1 - [1 - Q_4(t) - Q_5(t) + Q_4(t)\, Q_5(t)]$$

$$= Q_4(t) + Q_5(t) - Q_4(t)\, Q_5(t)$$

$Q_{12}(t)$ and $Q_{45}(t)$ are in series. Therefore,

$$R_{12}(t) = 1 - Q_{12}(t)$$

$$R_{45}(t) = 1 - Q_{45}(t)$$

$$\therefore R_{1245}(t) = R_{12}(t)\, R_{45}(t)$$

$$= \{1 - Q_{12}(t)\}\{1 - Q_{45}(t)\}$$

$$= 1 - Q_{12}(t) - Q_{45}(t) + Q_{12}(t)\, Q_{45}(t)$$

$$\therefore Q_{1245}(t) = 1 - R_{1245}(t)$$

$$= Q_{12}(t) + Q_{45}(t) - Q_{12}(t)\, Q_{45}(t)$$

Then, the equivalent diagram (Figure 5.26) becoming as follows:

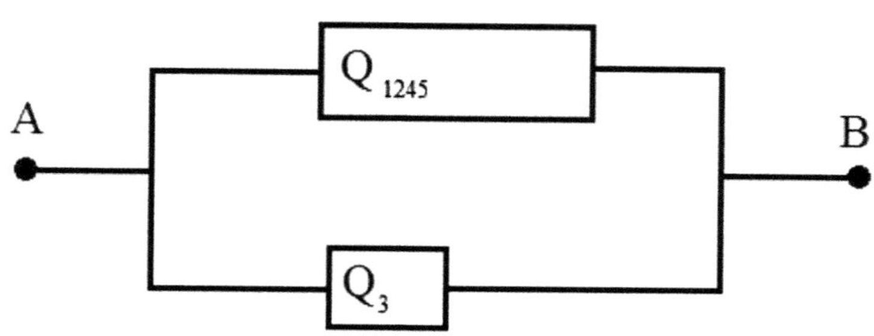

FIGURE 5.26 Equivalent diagram of non-identical components of a power system network.

$$\therefore Q_{12345}(t) = Q_{1245}(t)\, Q_3(t)$$
$$= [Q_{12}(t) + Q_{45}(t) - Q_{12}(t)\, Q_{45}(t)]\, Q_3(t)$$
$$= Q_{12}(t)\, Q_3(t) + Q_3(t)\, Q_{45}(t) - Q_{12}(t)\, Q_3(t)\, Q_{45}(t)$$

Therefore,

$$R_T(t) = 1 - Q_{12345}(t)$$
$$= 1 - Q_{12}(t)\, Q_3(t) - Q_3(t)\, Q_{45}(t) + Q_{12}(t)\, Q_3(t)\, Q_{45}(t)$$

R_T (t) is the overall reliability formula between the points A and B for the network in terms of un-reliabilities.

Example (5.18):

A power plant contains five components. The first and second components are connected in series, which form system (A); the third and fourth components are connected in series and form a system (B). The resultant of both systems (A) and (B) are connected in parallel by forming one equivalent system known as (C). Then system (C) is connected in series with a fifth component. The overall plant information is prepared in Table 5.10.

TABLE 5.10 Example 5.18 Data

Component No.	Failure Rate Per Week
1	1×10^{-5}
2	10×10^{-5}
3	2×10^{-4}
4	5×10^{-5}
5	15×10^{-5}

Calculate the overall reliability of the power plant when its components are running for 1000 weeks.

Solution:

For simplicity, the reliability of each component will be calculated through the 1000 weeks-time.

	Failure Rate	t (Weeks)	Reliability (t)
R_1	1.00E–05	1000	0.990049834
R_2	1.00E–04	1000	0.904837418
R_3	2.00E–04	1000	0.818730753
R_4	5.00E–05	1000	0.951229425
R_5	1.50E–04	1000	0.860707976
$A=R_1*R_2$	0.89583414	Q_A	0.104165865
$B=R_3*R_4$	0.77880078	Q_B	0.221199217
R_5	0.86070798	Q5	0.139292024
$Q_A * Q_B$	0.02304141		
R_{AB}	0.97695859		
$R_{AB} * R_5$	0.84087605	R-overall	

Example (5.19):

An electric power network has three identical transmission lines (consider at least six decimal points for each answer):

i. If the transmission lines are connected in series for which the summation of their failure rates is 0.003 failures per year. Determine the reliability of the network for the time equal to 12 years.
ii. If the four transmission lines are connected in parallel, what is the reliability of the new network for a period of 60 years?
iii. After calculating the reliability of the parallel transmission lines (consider the answer of part II as the total reliability for part III). Consider that each transmission line has a failure rate of 0.0025 per year, and the three of them are connected in series. Calculate the mission time in hours.

Solution:

i. $R_T(t) = R_1(t).R_2(t).R_3(t)$

$$R_T(12) = e^{-\lambda.t}$$

where: $\lambda = \lambda_T = 0.003$ failure/year

$$\therefore R_T(12) = e^{-(0.003 \times 12)} = 0.9646403$$

ii. In case that the three transmission lines are connected in Parallel. Therefore,
The failure rate (λ) for one transmission line is:

$$\lambda = \frac{0.003}{3} = 0.001 \; per\ year$$

$$R(t = 60\ years) = 1 - [\![[1 - e^{-0.001x60}]^3]\!]$$
$$= 0.9998025$$

iii. The total reliability for part (III) = The total reliability for part (II) = 0.9998025

$$R_T(t) = R_1(t).R_2(t).R_3(t)$$
$$0.9998025 = e^{-0.001xt}.e^{-0.001xt}.e^{-0.001xt}$$

∴ The mission time (t) = 24 hours.

Example (5.20):

An electric power network has four identical transmission lines (consider at least six decimal points for each answer):

i. If the transmission lines are connected in series for which the summation of their failure rates is 0.008 failures per year. Determine the reliability of the network for the time equal to 15 years.
ii. If the four transmission lines are connected in parallel, what is the reliability of the new network for a period of 50 years?
iii. After calculating the reliability of the parallel transmission lines (consider the answer of part II as the total reliability for part III). Consider that each transmission line has a failure rate of 0.0035 per year, and the three of them are connected in series. Calculate the mission time in hours.

Solution:

i. Since the network has four identical transmission lines. Therefore,

$$R_T(t) = R_1(t).R_2(t).R_3(t).R_4(t)$$
$$R_T(15) = e^{-\lambda t}$$

where: $\lambda = \lambda_T = 0.008$ failure/year.

∴ Failure rate for each transmission line = 0.002 failure/year R_T(15 years) = $e^{-(4 \times 0.002 \times 15)} = 0.886920$

ii. In case that the three transmission lines are connected in parallel. Therefore, the failure rate (λ) for one transmission line is:

$$\lambda = \frac{0.008}{4} = 0.002 \; per\ year$$

$$R(t = 50\ years) = 1 - [[[1 - e^{-0.001x50}]^4]]$$

$$= 0.999918$$

iii. The total reliability for part (III) = The total reliability for part (II) = 0.999918.

$$R_T(t) = R_1(t).R_2(t).R_3(t).R_4(t)$$

$$0.999918 = e^{-0.0035xt}.e^{-0.0035xt}.e^{-0.0035xt}.e^{-0.0035xt}$$

∴ The mission time (t) = 0.005857 year.

Which is 0.005857 year multiplied by (365 days × 24 hours) divide by 1 year. Therefore, the mission time (t) = 51.310675 hours.

5.7 NORMAL DISTRIBUTION

The normal distribution is known as a bell curve. It is asymmetrical distribution, in which the class with the lowest frequency is at each end of the curve and with the highest frequency. The highest frequency will be at the center of the same curve (Pan et al., 2017).

Example (5.21):

The lighting department of a city installed 2000 electric lamps, each having an average life of 1000 burning hours with a standard deviation of 200 hours, how many lamps may be expected to fail between 900 and 1300 running hours?

Solution:

n = 2000 lamps

μ = 1000 hours

σ = 200 hours

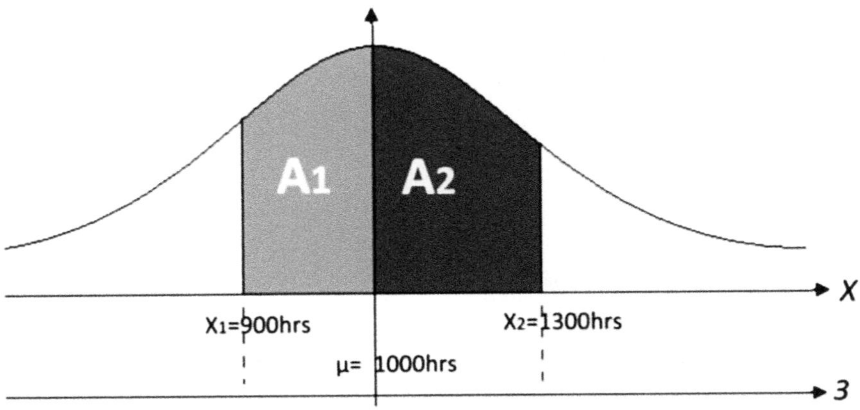

The number of lamps might be expected to fail between 900 and 1300 hours = ?

$$z_1 = \left|\frac{x_1 - \mu}{\sigma}\right| = \left|\frac{900 - 1000}{200}\right| = 0.5$$

Since 3_1 = 0.5, then the area under the curve can be obtained from the Appendix (B).

$A_1 = 0.1915$

$$z_1 = \left|\frac{x_2 - \mu}{\sigma}\right| = \left|\frac{1300 - 1000}{200}\right| = 1.5$$

Since 3_2 = 1.5, then the area under the curve can be obtained from the Appendix (B).

$A_2 = 0.4332$

Total area = $A_1 + A_2 = 0.1915 + 0.4332 = 0.6247$.

Expected number of lamps might fail = n (total area)

$= 2000 \times 0.6247 = 1249.4 \approx 1250$ lamps.

Example (5.22):

The lighting department of a city installed 2000 electric lamps, each having an average life of 1000 burning hours with a standard deviation of 200 hours, how many lamps may fail for the first 800 hours.

Solution:

n = 2000 lamps

μ = 1000 hours

σ = 200 hours

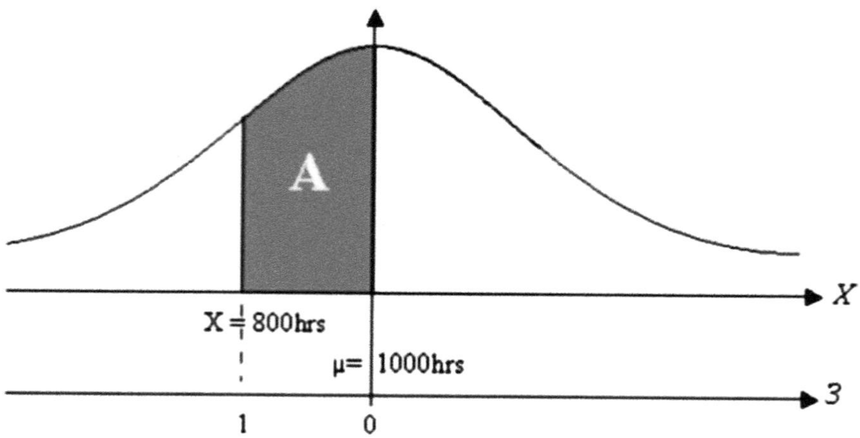

$$z_1 = \left|\frac{x_1 - \mu}{\sigma}\right| = \left|\frac{800 - 1000}{200}\right| = 1.0$$

Since 3 = 1.0, then the area under the curve can be obtained from the table.

$A_1 = 0.3413$

To find the area under the curve before 800 hours:

We have the curve, which is 100% as a full area; therefore, half of the curve will cover only 50% of the whole area. Then the area under the curve for the period before 800 hours will be 0.5000 – 0.3413 = 0.1587.

Expected number of lamps might fail = n (total area)

$= 2000 \times 0.1587 = 317.4 \approx 318$ lamps.

Example (5.23):

The lighting department of a city installed 2000 electric lamps, each having an average life of 1000 burning hours with a standard deviation of 200 hours, how many lamps might fail for the first 1200 hours?

Solution:

n = 2000 lamps

μ = 1000 hours

σ = 200 hours

$$z_1 = \left|\frac{x_1 - \mu}{\sigma}\right| = \left|\frac{1200 - 1000}{200}\right| = 1.0$$

Since 3 = 1.0, then the area under the curve can be obtained from Appendix (B).

$A_1 = 0.3413$

To find the area under the curve before 1000 hours, we have the curve, which is 100% as a full area; therefore, half of the curve will cover only 50%. Then the area under the curve for the period before 1000 hours will be 0.5000.

Therefore, the total area under the curve for the first 1200 hours is

$0.5 + 0.3413 = 0.8413$

Expected number of lamps might fail = n (total area)

$= 2000 \times 0.8413 = 1682.6 \approx 1683$ lamps.

Example (5.24):

The lighting department of a city installed 2000 electric lamps, each having an average life of 1000 burning hours with a standard deviation of 200 hours, how many lamps may be expected to fail between 800 and 1200 running hours?

Solution:

n = 2000 lamps

μ = 1000 hours

σ = 200 hours

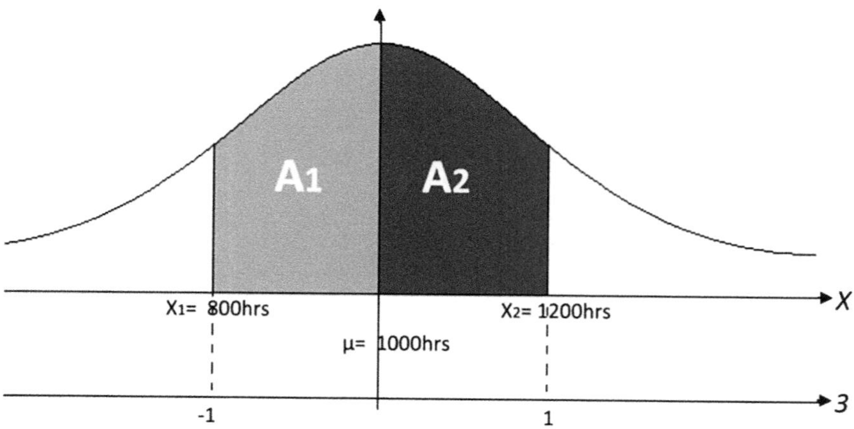

The number of lamps expected to fail between 800 and 1200 hours=?

$$z_1 = \left|\frac{x_1 - \mu}{\sigma}\right| = \left|\frac{800 - 1000}{200}\right| = 1.0$$

Since $3_1 = 1$, then the area under the curve can be obtained from Appendix (B).

$A_1 = 0.3413$

$$z_1 = \left|\frac{x_1 - \mu}{\sigma}\right| = \left|\frac{1200 - 1000}{200}\right| = 1.0$$

Since $3_2 = 1$, then the area under the curve can be obtained from the table.

$A_2 = 0.3413$

Total area $= A_1 + A_2 = 0.3413 + 0.3413 = 0.6826$

Expected number of lamps might fail = n (total area)

$= 2000 \times 0.6826 = 1365.2 \approx 1366$ lamps.

Example (5.25):

In one of the villages in the Kingdom of Bahrain, a number of electric lamps installed. A number of 1500 lamps expected to fail. From the normal distribution function table, two probabilities P_1 and P_2 needed. The ratio between both probabilities is represented by $P_2 = 1.926045\ P_1$. The number of burning hours will take place between 900 hours (for the first z) and 1540 (for the second z). Calculate z_1 and z_2, where z_1 has a negative value, and z_2 has a positive value. Then, determine both the standard deviation and the average life of the lamps. The probabilities summation is 0.4531722. Finally, calculate the total installed lamps.

Solution:

Total number of lamps expected to fail = 1500 lamps

$$P_2 = 1.926045\ P_1$$

Number of burning hours between 900 hours (for the first z) and 1540 (for the second z)

$$z_1(\text{negative}) = ? \text{ and } z_2(\text{positive}) = ?$$

$$\mu = ?$$

$$\sigma = ?$$

$$P_1 + P_2 = 0.4531722$$

$$P_1 = 0.4531722 - P_2$$

$$P_2 = 1.926045\ P_1$$

$$P_2 = 1.926045\ (0.4531722 - P_2)$$

$$P_2(1 + 1.926045) = 0.87283005$$

$$P_2 = 0.298297$$

From (Appendix B), $z_2 = 0.84$

$$P_1 = 0.4531722 - 0.298297$$

$$= 0.1548752$$

From (Appendix B), $z_1 = -0.4$

$$z_1 + z_2 = 0.4 + 0.84$$

$$\left|\frac{900-\mu}{\sigma} + \frac{\mu-1540}{\sigma}\right| = 1.24$$

$$\sigma = 516.129$$

and

$$\mu = 1106.45$$

Number of lamps expected to fail = 1500 lamps = n (total area)

$$= 1500 = n\,(0.298297 + 0.1548752)$$

$$n \approx 3310 \text{ lamps.}$$

This means that the number of burning hours between 900 hours (for the first z) and 1540 (for the second z) is approximately equal to 3310 Lamps.

Example (5.26):

The lighting department of a city installed 7000 electric lamps. Their average life is 1550 burning hours, with a standard deviation of 356 hours. Find the number of lamps that might be expected to fail in the last 500 burning hours.

Solution:

$$n = 7000 \text{ lamps}$$

$$\mu = 1550 \text{ hours}$$

$$\sigma = 356 \text{ hours}$$

The number of lamps might be expected to fail in the last 500 burning hours = ?

$$z = \left|\frac{x-\mu}{\sigma}\right| = \left|\frac{2600-1550}{356}\right| = 2.9494$$

$$\therefore \quad z \approx 2.95$$

From (Appendix B):

Probability = 0.4984.

The last 500 burning hours = 0.5 – 0.4984

$$= 0.0016$$

∴ The expected number of lamps might fail = (0.0016).(7000)

$$= 11.2$$

$$\approx 12 \text{ lamps.}$$

Example (5.27):

The lighting department in the Kingdom of Bahrain installed 5240 electric lamps in one of the villages. The electric lamp has an average life of 1950 burning hours with a standard deviation of 150 hours. Determine the burning hours for 2104 lamps expected to fail.

Solution:

n = 5240 lamps

μ = 1950 hours

σ = 150 hours

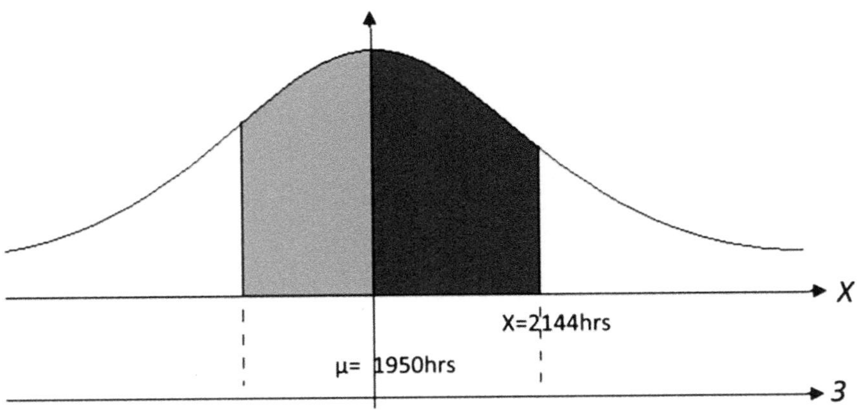

Number of lamps might be expected to fail = n × Area

2104 = (5240) × A

Therefore, A = 0.4015.

From the table: Since A=0.4015 Therefore, z = 1.29.

$$z = \left|\frac{x_1 - \mu}{\sigma}\right| = \left|\frac{1200 - 1000}{200}\right| = 1.0$$

$$1.29 = \left|\frac{x_1 - 1950}{150}\right|$$

Therefore,

$$x_1 - 1950 = 1.29 \times 150$$
$$x_1 = 2143.5$$
$$= 2144 \text{ hours.}$$

This means that it is expected to fail a number of 2104 lamps in the first 2144 hours.

Example (5.28):

The lighting department of a city installed 6000 electric lamps. Their average life is 1550 burning hours, with a standard deviation of 350 hours. In the last 500 burning hours, find the number of lamps that might be expected to fail.

Solution:

$n = 6000$ lamps

$\mu = 1550$ hours

$\sigma = 350$ hours

The number of lamps might be expected to fail in the last 500 burning hours = ?

$$x = (2 \times \text{Average Life Time}) - 500$$
$$\therefore x = (2 \times 1550) - 500$$

Then, $x = (2 \times 1550) - 500$

$$\therefore x = 3000 - 500 = 2600$$

$$z = \left|\frac{x-\mu}{\sigma}\right| = \left|\frac{2600-1550}{350}\right| = 3.0$$

$$\therefore \quad z \approx 3.0$$

From Appendix B:

Probability = 0.4986

The last 500 burning hours = 0.5 – 0.4986

= 0.0014

Therefore, the expected number of lamps might fail = (0.0014).(6000)

= 8.4

≈ 9 Lamps

5.8 BINOMIAL DISTRIBUTION

A binomial distribution has only two outcomes, and these outcomes are (p) and (q). The symbol (p) called the probability of success, where the symbol (q) defined as the probability of fail. Therefore, the expected outcome called success, and any other outcome is a failure. The probability of a failure (q) is equal to (1 – p). In other words, frequency distribution only two (mutually exclusive) outcomes are possible, such as better or worse, success or failure. This is in the language of electric power systems. Therefore, if the probability of success in any given trial is recognized, the binomial distributions can work to compute a given number of successes in a given number of trials. The probability of occurrence presented by the following equation (Billinton and Allan, 1992; Biscarri et al., 2018; Shonkwiler, 2016):

$$\text{Probability (Occurrence)} = \frac{n!}{r!(n-r)!}\, p^{r}\, q^{(n-r)}$$

$$p + q = 1$$

$$p + FOR = 1$$

where:

n: number of units

r: number of success

(n–r): number of failures

p: the probability of success

q: the probability of fail = λ = FOR

FOR: Forced Outage Rate

The FOR is used to define a case of the unit to be unavailable. The FOR can be also defined in a formula shape as:

$$FOR = \frac{\textit{Forced Outage Hours}}{\textit{Forced Outage Hours} + \textit{IN Service Hours}}$$

Example (5.29):

A small power network having five generators, each generator has 50MW capacity. Using the Binomial Distribution, find the cumulative probability for each state that the system will pass through. The FOR-for each generator is 0.02 per year.

Solution:

n = 5

Capacity of Generator = 50MW

FOR = 0.02 = λ = q

p = 1 – q = 1 – 0.02 = 0.98

Individual Probability = P (occurrence)

$$P(\text{Occurrence}) = \frac{n!}{r!(n-r)!}\, p^r\, q^{(n-r)}$$

No. of Success (r)	Capacity In (MW)	Capacity Out (MW)	Individual Probability	Cumulative Probability
5	250	0	0.9039	1.000
4	200	50	0.0922	0.09604
3	150	100	3.764×10^{-3}	3.842×10^{-3}
2	100	150	7.683×10^{-5}	7.761×10^{-5}
1	50	200	784×10^{-9}	787.2×10^{-9}
0	0	250	3.2×10^{-9}	3.2×10^{-9}

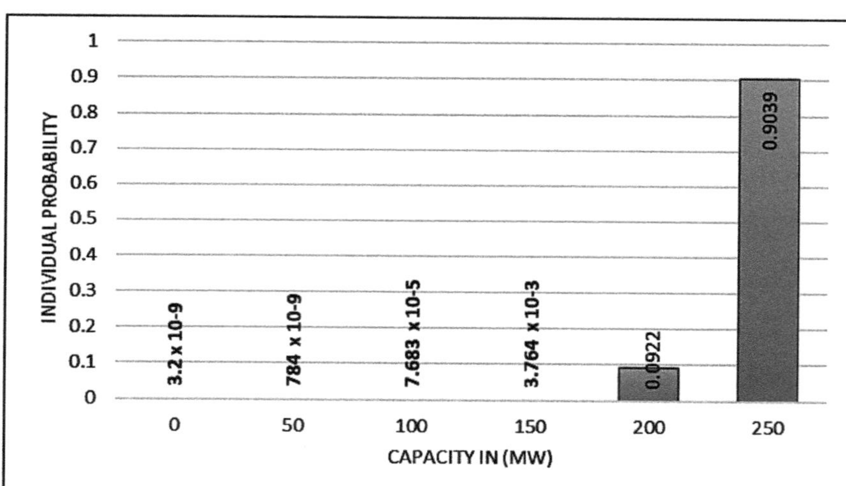

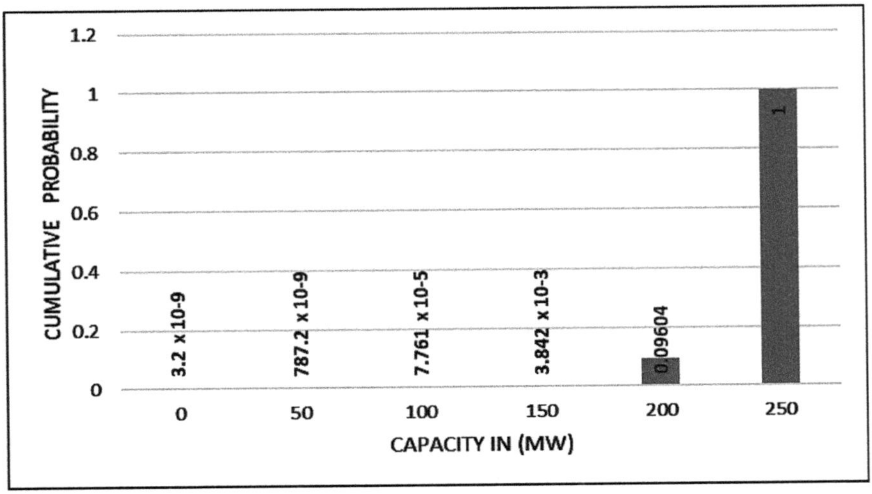

Example (5.30):

A power network has three generators, where each generator has 100MW capacity. Calculate the cumulative probabilities for each state that the system will pass through using the binomial distribution. The FOR-for each generator is 0.06 per year, show the capacities OUT and IN.

Solution:

n = 3 generators

Capacity of each Generator = 100 MW

FOR = 0.06 = λ = q

p = 1 – q = 1 – 0.06 = 0.94

Individual Probability = P(occurrence)

$$P(\text{Occurrence}) = \frac{n!}{r!(n-r)!}\ p^r\ q^{(n-r)}$$

No. of Success (r)	Capacity In (MW)	Capacity Out (MW)	Individual Probability	Cumulative Probability
3	300	0	0.830584	1.000
2	200	100	0.159048	0.169416
1	100	200	0.010152	0.010368
0	0	300	0.000216	0.000216

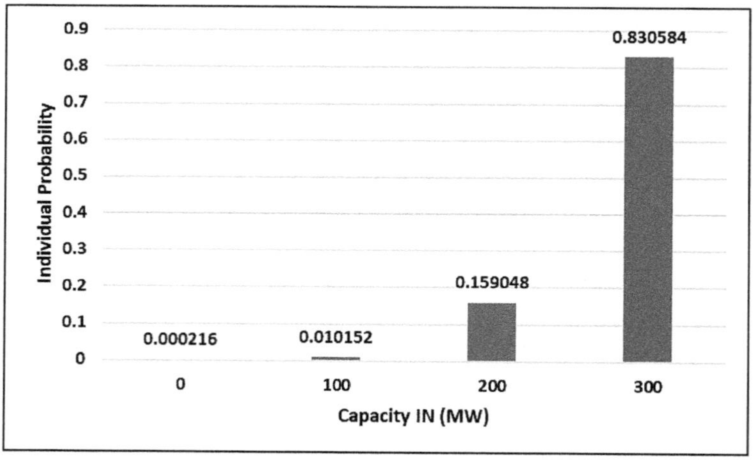

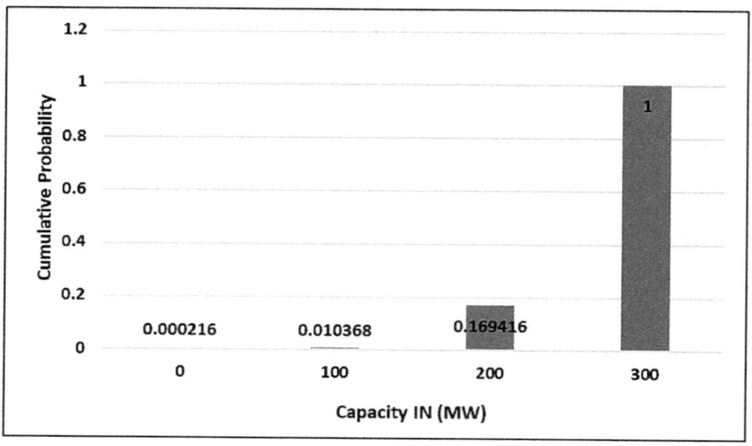

KEYWORDS

- **binomial distribution**
- **parallel components**
- **reliability block diagram**
- **series components**
- **series-parallel components**
- **un-repairable system**

REFERENCES

Almuhaini, M., & Al-Sakkaf, A. Markovian model for reliability assessment of microgrids considering load transfer. *Turkish Journal of Electrical Engineering and Computer Sciences*, **2017**, *25*, 4657–4672.

Al-Shaalan, A. M. (Accepted). Contingency selection and ranking for composite power system reliability evaluation. *Journal of King Saud University, Engineering Sciences*. Available Online November **2018**.

Billinton, R., & Allan, R. N. *Reliability Evaluation of Engineering Systems: Concepts and Techniques* (2nd edn.). Pitman: New York, **1992**.

Biscarri, W., Zhao, S. D., & Brunner, R. J. A simple and fast method for computing the Poisson binomial distribution function. *Computational Statistics and Data Analysis*, **2018**, *122*, 92–100.

Bollen, M. H. J. Understanding power quality problems: Voltage sags and interruptions." *IEEE Series on Power Engineering*, John Wiley & Sons, inc. Publication, **2000**.

Davidov, S., & Pantos, M. Optimization model for charging infrastructure planning with electric power system reliability check. *Energy,* **2019**, *166*, 886–894.

El-Hay, E. A., El-Hameed, M. A., & El-Fergany, A. A. Optimized parameters of SOFC for steady-state and transient simulations using interior search algorithm. *Energy*, **2019**, *166*, 451–461.

Feizabadi, M., & Jahromi, A. E. A new model for reliability optimization of series-parallel systems with non-homogeneous components. *Reliability Engineering and System Safety*, **2018**, *157*, 101–112.

Feng, D., Lin, S., Sun, X., & He, Z. Reliability assessment for traction power supply system based on quantification of margins and uncertainties. *Microelectronics Reliability*, **2018**, *88–90*, 1195–1200.

Groß, D., Arghir, C., & Dörfler, F. On the steady-state behavior of a nonlinear power system model. *Automatica*, **2018**, *90*, 248–254.

He, Y., & Zheng, Y. Short-term power load probability density forecasting based on Yeo-Johnson transformation quantile regression and Gaussian kernel. *Energy*, **2018**, *154*, 143–156.

IEEE Standard 493–2007. *IEEE Recommended Practice for the Design of Reliable Industrial and Commercial Power Systems: Gold Book*, **2007**.

Kim, M. C. Reliability block diagram with general gates and its application to system reliability analysis. *Annals of Nuclear Energy*, **2011**, *38*, 2456–2461.

Levitin, G., Jia, H., Ding, Y., Song, Y., & Dai, Y. Reliability of multi-state systems with free access to repairable standby elements. *Reliability Engineering and System Safety*, **2017**, *167*, 192–197.

Li, X., Huang, H., & Li, Y. Reliability analysis of phased mission system with non-exponential and partially repairable components. *Reliability Engineering and System Safety*, **2018a**, *175*, 119–127.

Li, X., Huang, H., Li, Y., & Zio, E. Reliability assessment of multi-state phased mission system with non-repairable multi-state components. *Applied Mathematical Modeling*, **2018b**, *61*, 181–199.

Mahmood, F., Hu, H., & Cao, L. Dynamic response analysis of corrosion products activity under steady state operation and mechanical shim based power-maneuvering transients in AP-1000. *Annals of Nuclear Energy*, **2018**, *115*, 16–26.

Mo, Y., Liu, Y., & Cui, L. Performability analysis of multi-state series-parallel systems with heterogeneous components. *Reliability Engineering and System Safety*, **2018**, *171*, 48–56.

Ota, S., & Kimura, M. A statistical dependent failure detection method for n-component parallel systems. *Reliability Engineering and System Safety*, **2017**, *167*, 376–382.

Pan, D., Liu, J., Huang, F., Cao, J., & Alsaedi, A. A Wiener process model with truncated normal distribution for reliability analysis. *Applied Mathematical Modeling*, **2017**, *50*, 333–346.

Pérez-Ràfols, F., & Almqvist, A. Generating randomly rough surfaces with given height probability distribution and power spectrum. *Tribology International*, **2019**, *131*, 591–604.

Qingqing, Z., Liudong, X., Rui, P., & Jun, Y. Aggregated combinatorial reliability model for non-repairable parallel phased-mission systems. *Reliability Engineering and System Safety*, **2018**, *176*, 242–250.

Shonkwiler, J. S. Variance of the truncated negative binomial distribution. *Journal of Econometrics*, **2016**, *195*, 209–210.

Zhao, P., Zhang, Y., & Chen, J. Optimal allocation policy of one redundancy in a n-component series system. *European Journal of Operational Research*, **2017**, *257*, 656–668.

CHAPTER 6

Generating Systems Reliability Indices

6.1 INTRODUCTION

The reliability studies depend on a number of indices. These indices are defined in the previous chapters and in the coming chapters (Kim, 2011; Rusin and Wojaczek, 2015). In the present chapter, a number of indices explained. The present chapter highlight on loss of load probability (LOLP), expected un-served energy (EUE), loss of load hours (LOLH), loss of load events (LOLEV), expected power not supplied (EPNS), loss of energy probability (LOEP), energy index of reliability (EIR), interruption duration index (IDI), expected energy not supplied (EENS), energy curtailed, loss of load expectation (LOLE), derated state, and forced outage rate (FOR).

6.2 PROBABILISTIC INDICES

The electric power system is the most likely complex system. The system divided into functional areas. These areas called generation, transmission, and distribution. To assess the system reliability, each part to be analyzed separately for easier evaluation and eventually by combining them together. Throughout the period, the resource adequacy metrics explain the occurrence of the risk, which helps the assessment of the power plant. The aggregation of risk across the year of operation might be more suitable to indicate the period of occurrence for resource adequacy risks within the area of operation (IEEE Standard 493–2007).

6.2.1 LOSS OF LOAD PROBABILITY (LOLP)

To show the quality and performance of an electrical system, the index LOLP needed. This index value is affected by load growth, the load duration curve, FOR of the plant, number, and capacity of generating units. This index defined as the probability of system as hourly or daily demand exceeding the available generating capacity within a given period. With a daily peak load curve, the LOLP can be calculated. The LOLP is a projected value of how much time the load on a power system is expected to be greater than the capacity of the available generating resources in the end. Therefore, the LOLP is based on combining the probability of generation capacity states with the daily peak probability. Then, as to assess the number of days during the year in which the generation system may be unable to meet the daily peak (Al-Shaalan, 2012; IEEE Standard 493–2007).

There are a number of problems that might arise during the use of LOLP of any power system reliability study. These problems are:

a. The LOLP does not provide any indication of the duration of shortages or frequency.
b. There are different techniques used to find LOLP. These techniques might obtain a different result for the same system.
c. The LOLP does not include additional emergency support that one control area may receive from another, even though any region does not include the additional emergency.
d. As a result of a future event not modeled by traditional LOLP calculation, major loss of load incidents usually occur.
e. The use of the LOLP index would be sufficient for the same system to investigate different expansion plans and an annual maintenance schedule. This is an argue statement and at the same time, it is correct only in case the duration peak load is static over any number of years of the study.
f. In the case that the utilities have different shapes of load curve, it is not useful to compare the reliabilities of utilities.
g. In most often the cumulative curve of any daily peak load, the used load model, which is used in the loss of load method, is not recognized in that model.
h. In the case of building the generation or signing purchasing power system equipment agreement, the LOLP is not important as an accurate predictor of the resulting incidence of electricity shortage.
i. The LOLP indicates the number of days in a year that might be lost. This indicates that the generation system is not able to meet the required load during those days.

6.2.2 EXPECTED UN-SERVED ENERGY (EUE) AND LOSS OF LOAD HOURS (LOLH)

This term belongs to the expected unserved energy. This index measures the expected amount of energy that failed to supply during the period of observation of the system load cycle due to shortages of basic energy supplies. In another way, it can be defined as a measure of the resource availability to serve the required load continuously at delivery points. It can be defined as the expected energy required that will not serve in a given year. The EUE can provide a measure relative to the size of a given assessment area (Billinton and Allan, 1996; IEEE Standard 493–2007).

At the same time, the expected number of hours per year can be known as the LOLH when a system's hourly demand is exceeding the generating capacity. This index can be calculated from the load duration curve. The hourly LOLE is known as LOLH when considering the latest case.

The LOLH has a number of classifications known as annual LOLH, monthly LOLH, and annual EUE, which is known as both actual and at the same time, normalized.

6.2.3 LOSS OF LOAD EVENTS (LOLEV)

In a case that some system load is not served during a period of time annually, this case is called LOLEV (NERC, 2016). The event might be an hour or a number of hours. At the same time, this is defined as the number of events where some system load is not served in through the year. A LOLEV can last for one hour or for several number of hours and can involve the loss of one or several hundred megawatts of load. This is known as the frequency of occurrence index (Billinton and Allan, 1996).

6.2.4 EXPECTED POWER NOT SUPPLIED (EPNS)

The EPNS (Elsaiah et al., 2015) is equivalent to the required curtailment since the problem aims to minimize the total load curtailment (Billinton and Allan, 1996). The EPNS is estimated as

$$EPNS = \sum_{i=1}^{n} LOL(i) \text{ x } Probability(i)$$

where: i is the number of cases; LOL is the loss of load.

6.2.5 LOSS OF ENERGY PROBABILITY (LOEP)

Loss-of-energy probability is a measure for generation reliability assessment (Gold Book, 2007). As a relationship, the LOEP is the ratio of the expected energy not served (EENS) during a certain long period of observation to the total energy demand during the same period (Billinton and Allan, 1996). Using the load duration curve is helping to calculate the LOEP for installed capacity. Therefore, the LOEP is becoming as

$$LOEP = \sum_i \frac{E_i \, . \, P_i}{E}$$

where the LOEP is unitless and it is a probability; Ei is the energy and it is not supplied due to a capacity outage (O_i); P_i is the probability of the capacity outage (O_i); E during the period of study, and E is the total energy demand.

6.2.6 ENERGY INDEX OF RELIABILITY (EIR)

The EIR (Venkatamuni and Reddy, 2014) is given by:

$$Energy\ Index\ of\ Reliability(EIR) = 1 - \frac{EENS}{Energy\ Demand}$$

where EENS is the expected energy not supplied (Billinton and Allan, 1996).

6.2.7 INTERRUPTION DURATION INDEX (IDI)

The IDI (Florencias-Oliveros et al., 2018) and (Indiana Utility Regulatory Commission Report) is defined as the IDI (Billinton and Allan, 1996). At the same time, the customer average interruption duration index (CAIDI).

$$Customer\ Average\ Interruption\ Duration\ Index(CAIDI) = \frac{System\ Average\ Interruption\ Duration\ Index(SAIDI)}{System\ Average\ Interruption\ Frequency\ Index(SAIFI)}$$

$$System\ Average\ Interruption\ Duration\ Index(SAIDI) = \frac{sum\ of\ all\ customer\ interruption\ durations}{total\ number\ of\ customers}$$

$$System\ Average\ Interruption\ Frequency\ Index(SAIFI) = \frac{total\ number\ of\ customer\ interruptions}{total\ number\ of\ customers}$$

6.2.8 EXPECTED ENERGY NOT SUPPLIED (EENS)

The term EENS (or Not Served) is the energy, which is needed to satisfy the load based on the generating units available. In other words, when the load exceeds the available generation, it is called EENS. This means that the energy is reflecting risk. The suitable unit used for the EENS is MWhr per year (Al-Shaalan, 2012; Billinton and Allan, 1996).

Example (6.1):

Consider the load duration curve shown below. This curve represents a variation of the load duration curve (Figure 6.1) of a small power network.

Three units are running to satisfy the above load curve, where the capacities of the units are:

Unit 1 ⟶ 50 MW

Unit 2 ⟶ 150 MW

Unit 3 ⟶ 300 MW

Find the EENS for the three periods shown on the curve.

Solution:

Period #1:

Energy = Area under the curve

$= (0.5 \times 5 \times 150) + (300 \times 5) = 1875$ MW.hr

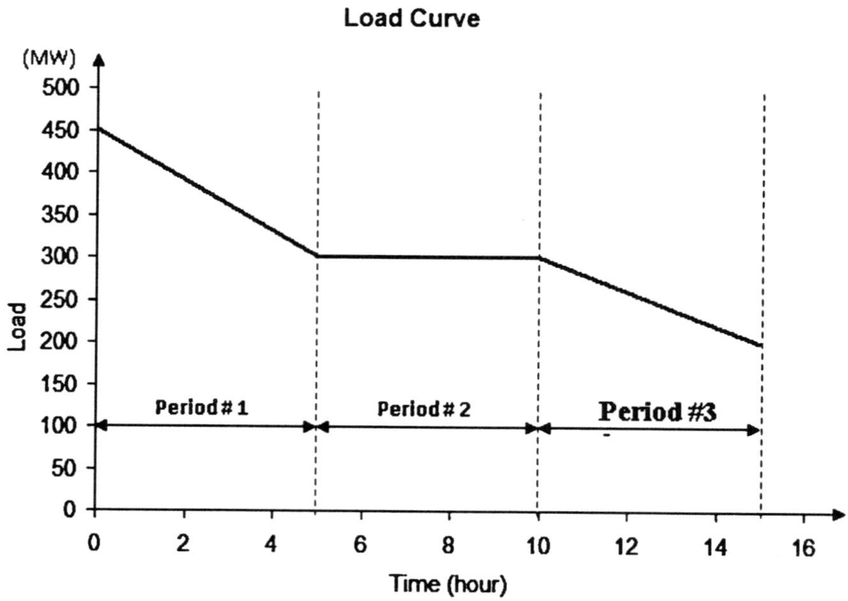

FIGURE 6.1 Load curve for the example 6.1.

To satisfy the first period presented in the curve, the generating-units (2 and 3) are running. Running the second and the third units with their full capacities will keep the network under risk. For this reason, the three units should be running.

Period #2:

To be in the safe side, the units 1 and 3 can be run together (50MW, 300 MW)

Energy = Area under the curve

$$= (300 \times 5) = 1500 \text{ MW.hr}$$

Period #3:

Units 1 and 3 can be run together (50 MW, 300 MW)

Energy = Area under the curve

$$= (0.5 \times 5 \times 100) + (5 \times 200) = 1250 \text{ MW.hr}$$

If all units (1, 2, and 3) are off, then the EENS will be the total area under the curve.

EENS = 1875 + 1500 + 1250 = 4625 MW.hr

Unit #	Period #		
	1	2	3
1	OFF	ON	ON
2	ON	OFF	OFF
3	ON	ON	ON

6.2.9 ENERGY CURTAILED

Due to a shortage in energy where the reason behind is that the demand for electricity exceeds the available power is known as curtailment. Therefore, the term curtailment is one among many tools to maintain system energy balance. It should be known that there are hierarchies of customers. This type of hierarchy is clear where some might require a partial or total cut back of electricity before others. The industrial sectors (industrial customers) are usually curtailed before reducing the service to the residential customers. When electrical energy service has to be interrupted, it means that energy reserves have dropped or are expected to drop below a certain level (Billinton and Allan, 1996; Denholm and Mai, 2019; IEEE Standard 493–2007; Schermeyer et al., 2018).

$$\text{Energy Curtailed} = \text{Total Energy} - \text{Total Time} \times \text{Capacity IN}$$

$$\text{Expectation of Energy Curtailed} = \text{Probability} \times \text{Energy Curtailed}$$

Example (6.2):

Consider a small electric power system network represented by a load duration curve shown in Figure 6.2. The load curve is represented for a period of 100 hours. The network is running by a generating unit through three-states shown in Table 6.1. Calculate the Energy Curtailed for the considered case. Then, find the generating-unit expected produced energy.

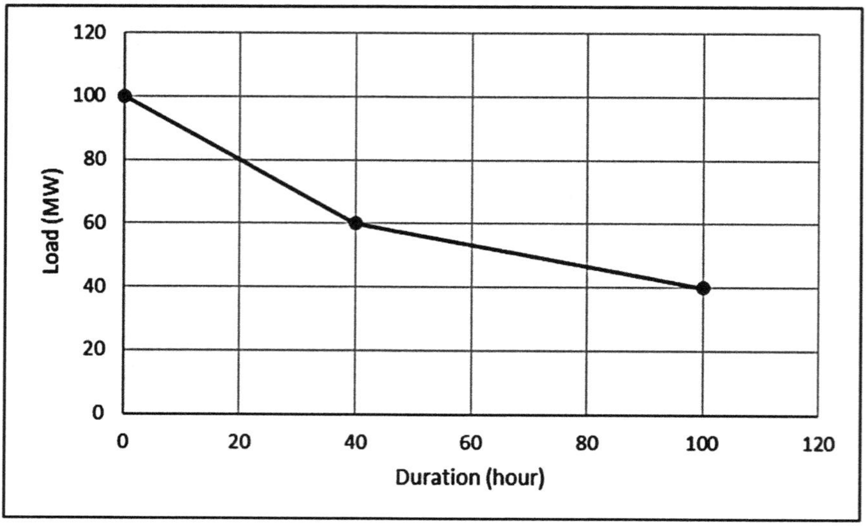

FIGURE 6.2 Load-duration curve.

TABLE 6.1 Three-States for Generating-Unit

Capacity IN (MW)	Capacity OUT (MW)	Probability
50	0	0.65
30	10	0.30
0	25	0.05

Solution:

First, the total energy under the curve is calculated as:

$$\text{Energy} = [0.5 \times 40 \times (100 - 60)] + [0.5 \times 60 \times (60 - 40)] + [60 \times 40] + [40 \times 60]$$

$$\text{Energy} = 800 + 600 + 2400 + 2400$$

$$\text{Energy} = 6200 \text{ MWhr}$$

Therefore, the expected energy produce by the unit-generator is the difference between energy under the curve ($EENS_0$, the generating unit or units is/are OFF) and the summation of the expectation of energy curtailed. This means that the generating-unit expected produced energy is 4150 MWhr.

Capacity IN (MW)	Capacity OUT (MW)	Probability	Energy Curtailed (MWhr)	Expectation of E. C. (MWhr)
50	0	0.65	6200–(100×50) = 1200	0.65×1200 = 780
30	10	0.30	6200–(100×30) = 3200	0.30×3200 = 960
0	25	0.05	6200–(100×0) = 6200	0.05×6200 = 310
				Σ = 2050 MWhr

Example (6.3):

Consider a small electric power system network represented by a load duration curve shown in Figure 6.3. The load curve is represented for a period of 100 hours. The network is running by two-generating units through the three-state model and two-state model as shown in Tables 6.2 and 6.3. Calculate the energy curtailed for the considered case. Then, find the generating-units expected produced energy.

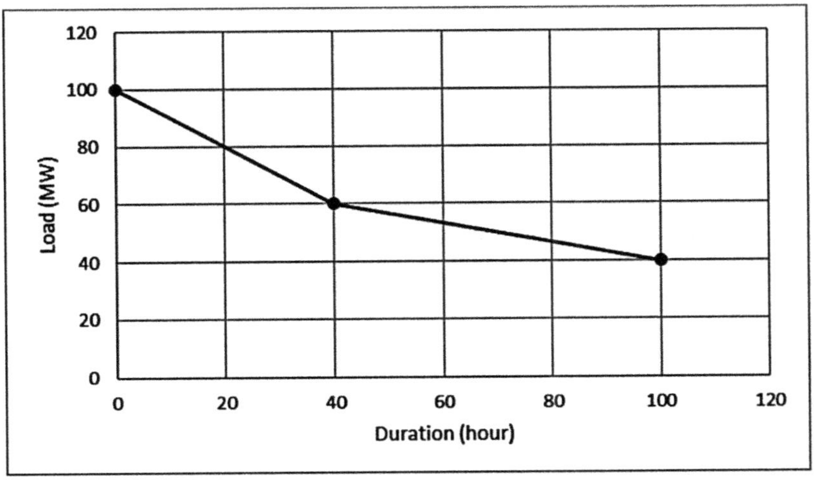

FIGURE 6.3 Small electric power network load-duration curve.

TABLE 6.2 Three-State Model Data

Cap. IN (MW)	Cap. OUT (MW)	Probability
25	0	0.65
15	10	0.3
0	25	0.05

TABLE 6.3 Two-State Model Data

Cap. IN (MW)	Cap. OUT (MW)	Probability
30	0	0.97
0	30	0.03

Solution:

First, the total energy under the curve is calculated as:

$$\text{Energy} = 6200 \text{ MWhr}$$

The results are shown in Tables 6.4 and 6.5.

TABLE 6.4 Three-State Model Results

Cap. IN (MW)	Cap. OUT (MW)	Prob.	Curtailed Energy	Expectation of E. C. (MWhr)
25	0	0.65	3700	2405
15	10	0.3	4700	1410
0	25	0.05	6200	310

The total expectation of energy curtailed = 4125 MWhr

TABLE 6.5 Two-State Model Results

Cap. IN (MW)	Cap. OUT (MW)	Prob.	Curtailed Energy (MWhr)	Expectation of E. C. (MWhr)
30	0	0.97	3200	3104
0	30	0.03	6200	186

The total expectation of energy curtailed = 3290 MWhr

Joining both generating-units together:

Cap. IN (MW)	Cap. OUT (MW)	Prob.	Energy Curtailed (MWhr)	Expectation (MWhr)	Connect
55	0	0.6305	700	441.35	25 + 30
45	10	0.291	1700	494.7	15 + 30
30	25	0.0485	3200	155.2	0 + 30
25	30	0.0195	3700	72.15	25 + 0
15	40	0.009	4700	42.3	15 + 0
0	55	0.0015	6200	9.3	0 + 0
		Σ 1		Σ 1215 MWhr	

6.2.10 LOSS OF LOAD EXPECTATION (LOLE)

The term LOLE is one of the reliability indices. During the observation of the power system load cycle, the expected number of hours (HLOLE: Hourly Loss of Load Expectation) or time period are calculated when insufficient generating capacity is available to serve the required load at a time. In another statement, it can be defined as the expected number of days per annual period in which the available generation capacity is insufficient to satisfy and serve the daily maximum load. It can be calculated from the daily peak load or annual peak load. In practical applications, the LOLE expectation index is more often used than the LOLP probability index. The relationship between both the LOLP and the LOLE is formulated as (IEEE Standard 493–2007):

$$LOLE = LOLP.T$$

In case the load model is an annual continuous load curve with day maximum load, then the LOLE unit is in days per year and T = 365 days.

In case the load model is an hourly load curve, then the LOLE unit is in hours per year and T = 8760 hours.

Example (6.4):

A power station is supplying a city in the Kingdom of Bahrain for a year. The weekly average loads have been recorded as such (in MW):

77	46	41	52
34	41	34	41
41	34	41	57
41	46	57	77
41	52	34	34
57	46	77	41
52	57	41	34
41	41	52	57
34	57	34	52
41	77	34	34
57	52	57	46
77	46	52	34
57	41	52	46

The power station supplying the city has three generating units, the first unit is working with a capacity of 25MW and 0.02 fail per week. The second unit has a capacity of 25 MW as well, with a failure rate equal to 0.05 f/week. The last unit (number 3) has a repair rate of 0.96 r/week, and its capacity is 50 MW. Calculate the LOLE in percentage.

Solution:

Weekly Average Load (MW)	77	57	52	46	41	34
No. of Occurrence (week)	5	9	8	6	13	11

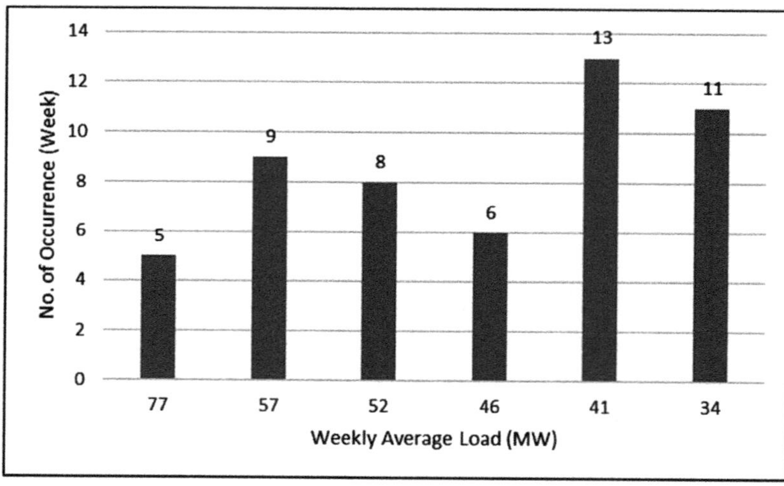

Unit #	Capacity (MW)	Prob. of Success	Prob. of Fail
1	25	0.98	0.02
2	25	0.95	0.05
3	50	0.96	0.04

Capacity IN (MW)	Capacity OUT (MW)	Ind. Probability	Cumulative Prob.
ON-ON-ON 100	0	(0.98)(0.95)(0.96) = 0.89376	0.10624 + 0.89376 = 1.00000
ON-OFF-ON OFF-ON-ON 75	25	(0.98)(0.05)(0.96)+ (0.02) (0.95)(0.96) = 0.06528	0.04096 + 0.06528 = 0.10624
OFF-OFF-ON ON-ON-OFF 50	50	(0.02)(0.05)(0.96)+ (0.98) (0.95)(0.04) = 0.0382	0.00276 + 0.0382 = 0.04096
ON-OFF-OFF OFF-ON-OFF 25	75	(0.98)(0.05)(0.04)+ (0.02) (0.95)(0.04) = 0.00272	0.00004 + 0.00272= 0.00276
OFF-OFF-OFF 0	100	(0.02)(0.05)(0.04) = 0.00004	0.00004

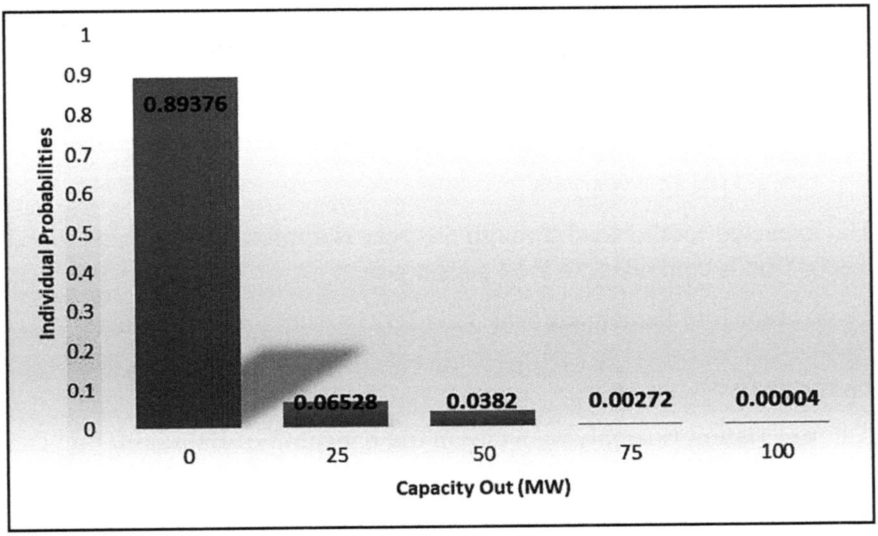

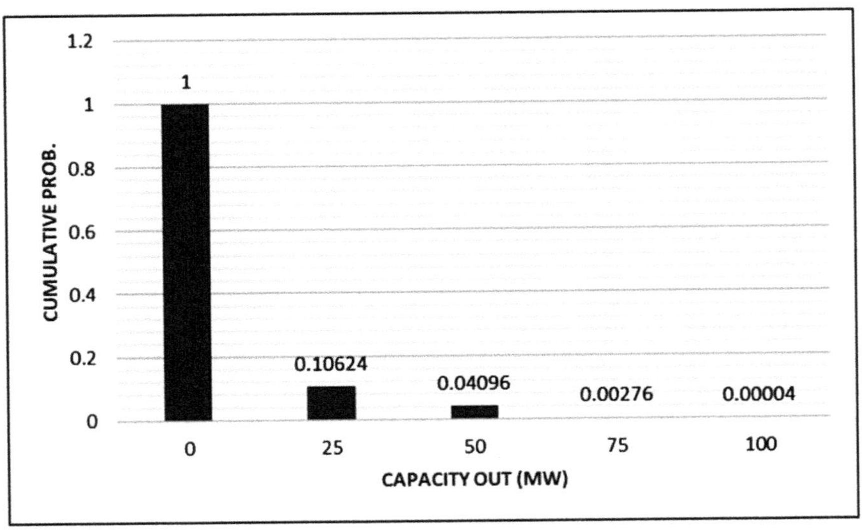

$$\text{LOLE} = \sum_{i=1}^{N} (\text{No. of Occurrences}) \times P_i (\text{Maximum Available Power}) - \text{Peak Load}$$

where: the maximum available power is the maximum capacity.

LOLE = 5 P(100 – 77) + 9 P(100 – 57) + 8 P(100 – 52) + 6 P(100 – 46) + 13 P(100 – 41) + 11 P(100 – 34)

$$= 5\ P(23) + 9\ P(43) + 8\ P(48) + 6\ P(54) + 13\ P(59) + 11\ P(66)$$

$$= 5(0.10624)+9(0.04096)+8(0.04096)+6(0.00276)+13(0.00276) + 11(0.00276)$$

$$= 0.5312 + 0.36864 + 0.32768 + 0.01656 + 0.03588 + 0.03036$$

$$= 1.310322 \text{ week/year}$$

The expected loss of load through the year is approximately equal to 1.3 week. This is equivalent to 9.17 days a year.

$$\text{LOLE in Percentage} = (1.31032/52) \times 100\% = 2.519846\%$$

Example (6.5):

A power station is supplying a city in the Kingdom of Bahrain for a year. The weekly average loads have been recorded as such (in MW):

68	92	82	104
82	114	82	114
92	82	154	68
114	154	114	68
82	114	68	114
82	68	154	104
68	82	114	82
104	82	154	114
114	104	82	92
68	82	104	82
82	68	154	104
92	104	68	92
68	104	92	68

The power station supplying the city has three generating units, the first unit is working with a capacity of 50MW and 0.02 fail per week. The second unit has a capacity of 50MW as well, with a failure rate equal to 0.05 f/week. The third unit has a repair rate of 0.96 r/week, and its capacity is 100MW. Calculate the LOLE in percentage.

Solution:

Weekly Average Load (MW)	68	82	92	104	114	154
No. of Occurrence (week)	11	13	6	8	9	5

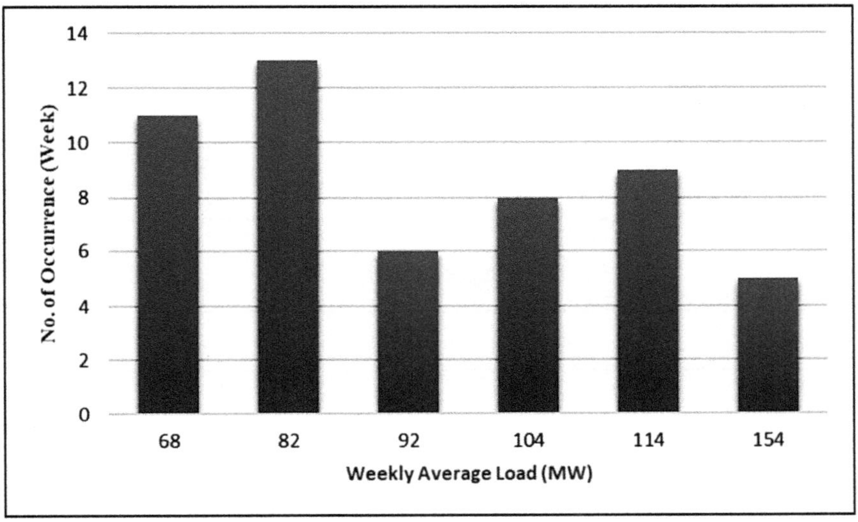

Unit #	**Capacity (MW)**	**Probability of Success**	**Probability of Fail**
1	50	0.98	0.02
2	50	0.95	0.05
3	100	0.96	0.04

Capacity IN (MW)	**Capacity OUT (MW)**	**Individual Probability**	**Cumulative Probability**
ON-ON-ON 200	0	(0.98)(0.95)(0.96) = 0.89376	0.10624 + 0.89376 = 1.00000
ON-OFF-ON OFF-ON-ON 150	50	(0.98)(0.05)(0.96)+ (0.02)(0.95)(0.96) = 0.06528	0.04096 + 0.06528= 0.10624
OFF-OFF-ON ON-ON-OFF 100	100	(0.02)(0.05)(0.96)+ (0.98)(0.95)(0.04) = 0.0382	0.00276 + 0.0382 = 0.04096
ON-OFF-OFF OFF-ON-OFF 50	150	(0.98)(0.05)(0.04)+ (0.02)(0.95)(0.04) = 0.00272	0.00004 + 0.00272= 0.00276
OFF-OFF-OFF 0	200	(0.02)(0.05)(0.04) = 0.00004	0.00004

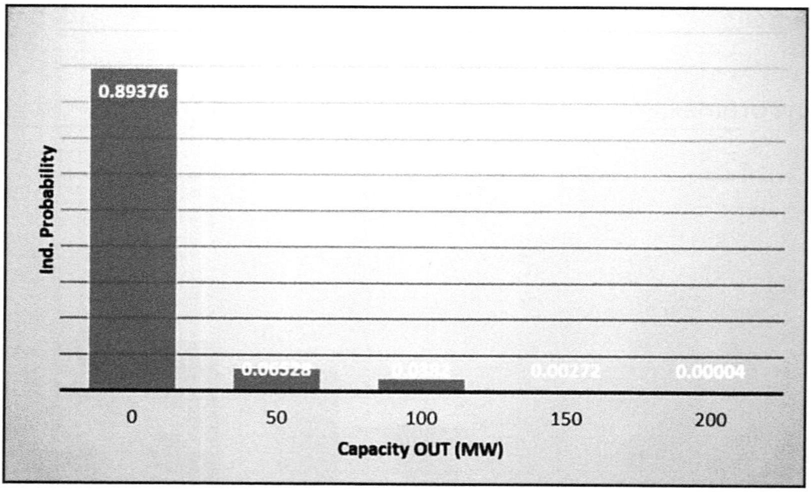

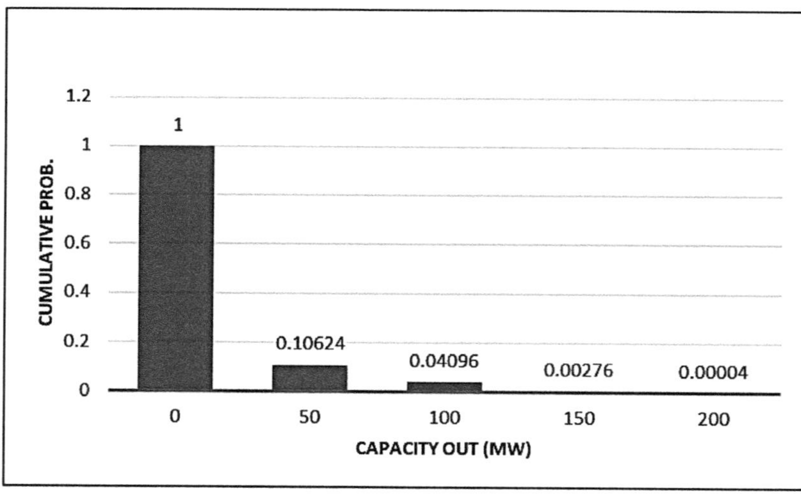

$$LOLE = \sum_{i=1}^{N} (No.\ of\ Occurrences)\, \mathrm{x} P_i\, (Maximum\ Available\ Power) - Peak\ Load$$

where: the maximum available power is the maximum capacity.

LOLE = 13 P(200 – 82) + 11 P(200 – 68) + 9P(200 – 114) + 8 P(200 – 104) + 6 P(200 – 92) + 5 P(200 – 154)

= 13 P(118) + 11 P(132) + 9 P(86) + 8 P(96) + 6 P(108) + 5 P(46)

= 13(0.00276) + 11 (0.00276) + 9 (0.04096) + 8 (0.04096) + 6 (0.00276) + 5 (0.10624)

= 1.310322 week/year

The expected loss of load through the year is approximately equal to 1.3 week. This is equivalent to 9.17 days a year.

LOLE in Percentage = (1.31032/52) × 100% = 2.519846%

Example (6.6):

Three generators have been running for 52 weeks. These generators are represented by the following data summarized in the following table:

Unit #	Capacity (MW)	Prob. of Success	Prob. of Fail
1	50	0.998	0.002
2	60	0.995	0.005
3	100	0.992	0.008

The table below shows the data of the peak loads it occurred and how frequent it is happening:

Weekly Average Load (MW)	68	82	92	104	114	154
No. of Occurrence (Week)	11	13	6	8	9	5

Calculate the LOLE in percentage.

Solution:

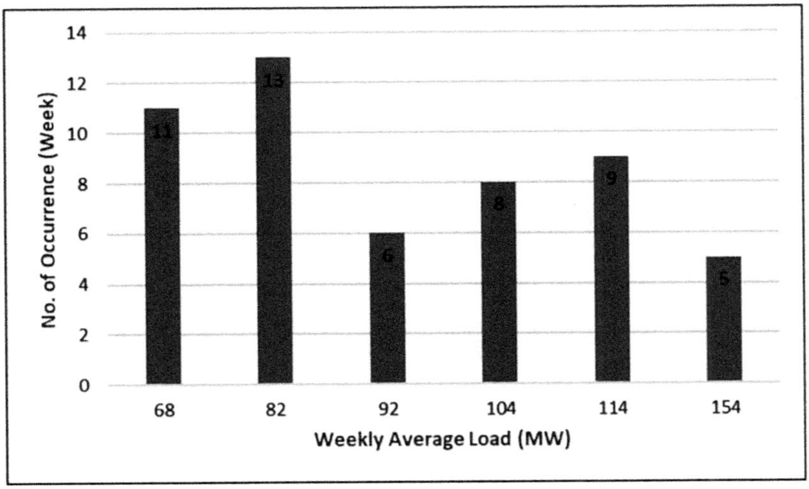

Capacity IN (MW)	Capacity OUT (MW)	Ind. Probability	Cumulative Prob.
ON-ON-ON 210	0	(0.998)(0.995)(0.992) = 0.98506592	1.00000000
OFF-ON-ON 160	50	(0.002)(0.995)(0.992) = 0.00197408	0.01493408
ON-OFF-ON 150	60	(0.998)(0.005)(0.992) = 0.00495008	0.01296
ON-ON-OFF 110	100	(0.998)(0.995)(0.008) = 0.00794408	0.00800992
OFF-OFF-ON 100	110	(0.002)(0.005)(0.992) = 0.00000992	0.00006584
OFF-ON-OFF 60	150	(0.002)(0.995)(0.008) = 0.00001592	0.00005592
ON-OFF-OFF 50	160	(0.998)(0.005)(0.008) = 0.00004	0.00004
OFF-OFF-OFF 0	210	(0.002)(0.005)(0.008) = 80×10^{-9}	80×10^{-9}

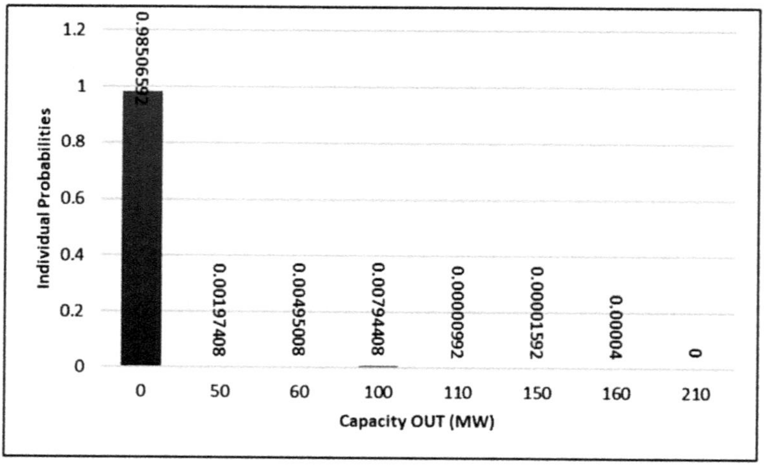

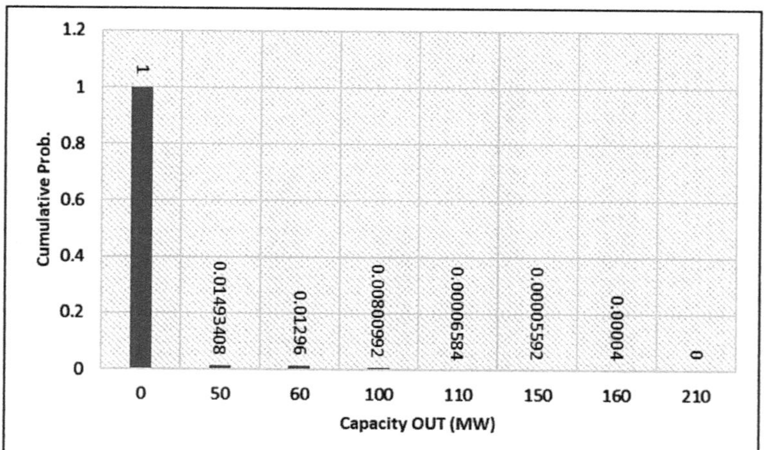

$$LOLE = \sum_{i=1}^{N} (No.\ of\ Occurrences) \text{x} P_i (Maximum\ Available\ Power) - Peak\ Load$$

where: the maximum available power is the maximum capacity.

LOLE = 7 P(210 – 200) + 11 P(210 – 160) + 10 P(210 – 125) + 13 P(210 – 95) + 5 P(210 – 80) + 6 P(210 – 75)

= 7 P(10) + 11 P(50) + 10 P(85) + 13 P(115) + 5 P(130) + 6 P(135)

= 7 (0.1493408) + 11 (0.1493408) + 10 (0.00800992) + 13 (0.00005592) + 5 (0.00005592)

+ 6 (0.00005592)

= 2.76957568 week/year

The expected loss of load through the year is approximately equal to 2.77 weeks. This is equivalent to 19.4 days a year.

LOLE in Percentage = (2.76957568/52) × 100% = 5.326107077%

Example (6.7):

Three generators are running with the following data:

Unit #	Capacity (MW)	λ (f/day)	μ(r/day)
1	25	0.02	0.98
2	25	0.03	0.97
3	50	0.04	0.96

The three generators have been running for one year (365 days). The data for a period of one year with a maximum of 100MW are given in the following table:

Daily Peak Load (MW)	57	52	46	41	34
No. of Occurrence (days)	12	83	107	116	47

Find the LOLE.

Solution:

Unit #	Probability of Success	Probability of Fail
1	0.98	0.02
2	0.97	0.03
3	0.96	0.04

To combine the three units together as one system:

Capacity In (MW)	Capacity Out (MW)	Individual Probability	Cumulative Probability
ON ON ON 100	0	(0.98)(0.97)(0.96) = 0.912576	0.912576 + 0.087424 = 1.000000
ON OFF ON OFF ON ON 75	25	(0.98)(0.03)(0.96)+(0.02)(0.97)(0.96) = 0.046848	0.046848 + 0.040576 = 0.087424
ON ON OFF OFF OFF ON 50	50	(0.98)(0.97)(0.04)+(0.02)(0.03)(0.96) = 0.038600	0.038600 + 0.001976 = 0.040576
ON OFF OFF OFF IFF ON 25	75	(0.98)(0.03)(0.04)+(0.02)(0.97)(0.04) = 0.019520	0.019520 + 0.000024 = 0.001976
OFF OFF OFF 0	100	(0.02)(0.03)(0.04) = 0.000024	0.000024

$$LOLE = \sum_{i=1}^{N} (No.\ of\ Occurrences)\, xP_i\,(Maximum\ Available\ Power) - Peak\ Load$$

where: the maximum available power is the maximum capacity.

Daily Peak Load (MW)	57	52	46	41	34
No. of Occurrence (days)	12	83	107	116	47

LOLE = 12×P(100 – 57) + 83×P(100 – 52) + 107×P(100 – 46)
+ 116×P(100 – 41) + 47×P(100 – 34)

LOLE = 12×P(43) + 83×P(48) + 107×P(54)+ 116×P(59) + 47×P(66)

LOLE = 12×(0.040576) + 83×(0.040576) + 107×(0.001976)
+ 116×(0.001976)+ 47×(0.001976)

LOLE = 4.38824 days/year

This figure shows that the system might loss power for 4.38824 days in a year. This means that the expected number of hours per year that the system's electricity production cannot meet its demand.

The loss of load percentage = (4.38824 / 365) ×100 = 1.2327%

6.2.11 DERATED STATE

Usually, the term derating is used in the electric power systems and electronic devices. At the same time, the derating as a term used when a component rated less than the maximum capability of it. In the field of electric power systems, the expected excess of available generation capacity over electric power demand is known as the derated capacity margin (Billinton and Allan, 1992; IEEE Standard 493–2007). The de-rated capacity margin statement is used; because uncontinuous plant generation depending on weather conditions, and at the same time, the plants provide uncertain levels of generation capacity. At a case of some generating-units at non-peak demand times are under maintenance, the system could be considered under the derated state. As an example, the three-state model is considered as shown in Figure 6.4. This model has the following states:

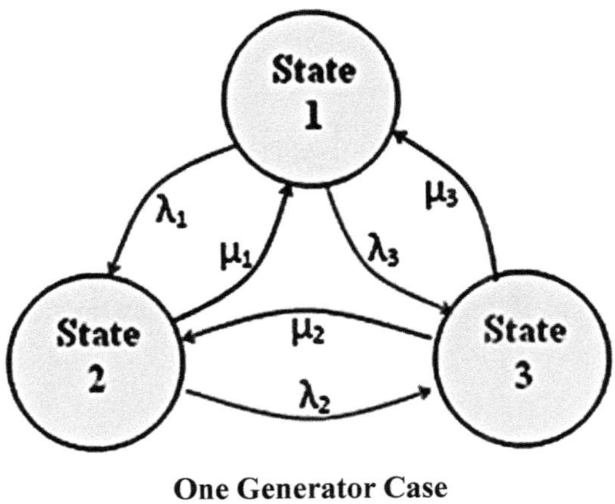

FIGURE 6.4 Three-state model with a derated-state.

- State (1): it means that the system is operated with full capacity.
- State (2): it means that the system is working with the partial of its capacity (which is defined as a derated-state).
- State (3): it means that the system is in the fail state.

Therefore,

$$P_1(t) + P_2(t) + P_3(t) = 1$$

6.2.12 FORCED OUTAGE RATE (FOR)

This index varies between the resource types and areas. The FOR can be also defined in a formula shape as (Billinton and Allan, 1992; IEEE Standard 493–2007):

$$FOR = \frac{Forced\ Outage\ Hours}{Forced\ Outage\ Hours + IN\ Service\ Hours}$$

6.2.13 AVAILABILITY AND UNAVAILABILITY

The availability (Davidov and Pantos, 2019) of the system has a number of formulas:

$$\text{Availability} = \frac{\text{MTBF}}{\text{MTBF} + \text{MTTR}}$$

At the same time, when the system has two-states the availability of the system is given by

$$\text{Availability of the system} = \text{Probability at any time when the system in the Up State}$$

$$\text{Unavailability} = \text{P}(\text{Down})$$

In case, the system has three states: up-state, derated-state, and down-state. Therefore,

$$\text{Availability} = \text{P}(\text{Up}) + \text{P}(\text{Derated})$$

$$\text{Unavailability} = \text{P}(\text{Down})$$

This means the availability is considered that the system is under operation. The unavailability means that the system is in the fail state.

6.3 GENERATION SYSTEM RELIABILITY EVALUATION (MEASUREMENT)

Any power system is evaluated based on a number of points, these points are summarized as (Akhavein and Porkar, 2017; Al-Shaalan, 2012; Chaiamarit and Nuchprayoon, 2013; Zhong et al., 2017):

a. The unit(s) availability during the operation time.
b. The unit(s) unavailability during the operation time.
c. The system components, which are classified as repairable or non-repairable.
d. The components might be in service or out of service.
e. The state probabilities need to known.
f. The load forecast during the electric network operation.

6.3.1 GENERATION AND LOAD MODELS

The generation model (Ma et al., 2018; Staveley-O'Carroll, 2019) is represented in its simplest form, which is a two-state model. This means that it will pass through the up-state and down-state. This case has been discussed earlier. The load model (Wang et al., 2018; Lv et al., 2018; Zheng et al., 2019) is represented by the load curve.

6.3.2 SYSTEM RISK INDICES

A combination of the generating capacity and the load models will result in the system risk indices. A random sequence of N number of load levels is known as a load cycle for the period (Mosaad et al., 2018; Yu et al., 2018). The N number of load periods must be an integer. As an example, if a daily load model has a peak load level of mean duration e day (e is known as the exposure factor) and a fixed low load of $(1 - e)$ day. The electric load exposure factor is represented in the impact of the risk over the load asset, or percentage of load asset lost. As an example, if the load asset value is reduced two thirds, the load exposure factor value is 0.667. If the asset is completely lost, the exposure factor is unity. Therefore, this case can be represented as a daily load model curve as illustrated in Figure 6.5.

With reference to Figure 6.5, the value of X is proportional inversely to the electric load exposure factor (e). This means that when the value of (X) decreases, (e) increases. The electric load cycle can be presented as shown in Figure 6.6.

The parameters needed to completely define the load cycle period shown in Figure 6.6 are summarized as shown in Table 6.6.

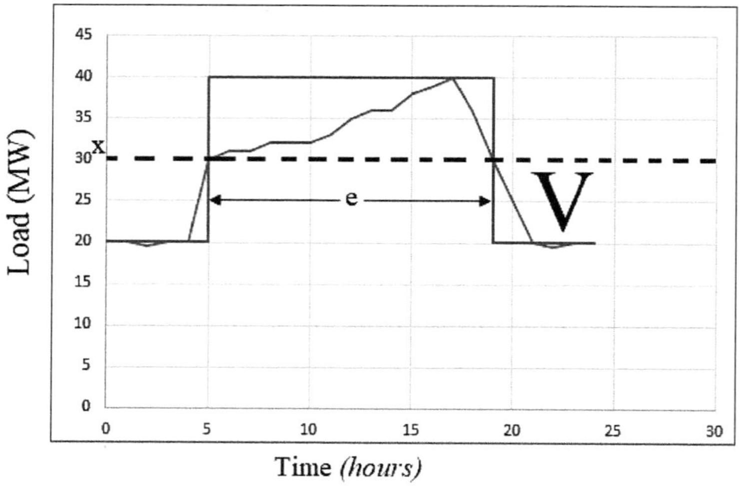

FIGURE 6.5 Daily electric load model.

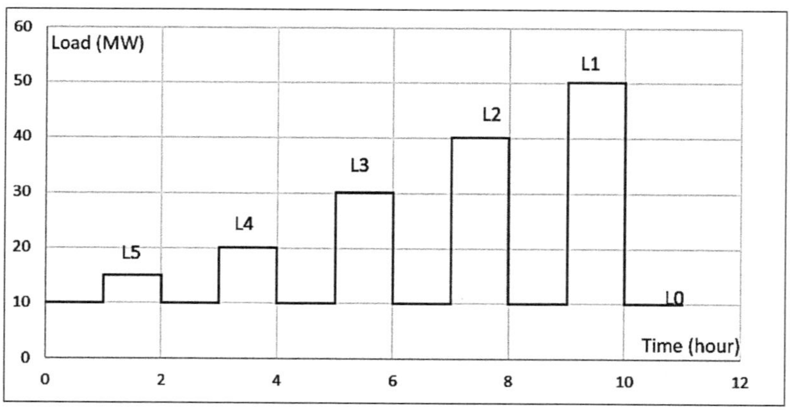

FIGURE 6.6 Load cycle period.

The combination of discrete levels, the capacity available of the electric system, and discrete levels of electric system demand or electric load form a set capacity margins (m_k). It is known that the margin is defined as the difference between the available capacity of the system and the system load. In case that the margin signs is becoming negative. Therefore, it can be concluded that the load is exceeding the available capacity. At this stage, the system might fail. The relationship of such condition can be written in the following form:

$$m_k = P_C - L_i$$

and

$$P_k = P_C . P_i$$

TABLE 6.6 Load Cycle Period Parameters, Load Model

Item	Parameter(s)
Number of Load Levels	N
Peak Electric Loads (MW)	L_i, i–1, ………, N such that $L_1 > L_2 > L_3 > L_4 > L_5 > \ldots\ldots > L_N$
Low Electric Load	L_0
Number of Occurrences of L_i	n_i, i = 1, ………, N
Interval Length, hours	$D = \sum_{i=1}^{N} n_i$
Mean Duration for the Peak Load	e
Mean Duration for the Low Load	1 – e
Transition Rate to Peak Load (Upward Load)	$\lambda_{+Li} = 0$
Transition Rate to Peak Load (Downward Load)	$\lambda_{-Li} = \frac{1}{e}$
Transition Rate to Low Load (Upward Load)	$\lambda_{+L0} = \frac{1}{1-e}$
Transition Rate to Low Load (Downward Load)	$\lambda_{-L0} = 0$
Frequency of occurrence of L_i	$f_i = \frac{n_i}{D}$
Frequency of occurrence of L_0	$f_0 = 1$
Probability of Peak Load	$P(L_i) = \frac{n_i}{D} e$
Probability of Low Load	$P(L_0) = 1 - e$

At the same time, departure rates can be written in the following form:

$$\lambda_{+m} = \lambda_{+C} + \lambda_{-L}$$

$$\lambda_{-m} = \lambda_{-C} + \lambda_{+L}$$

where: λ_{+m} is known as the upward margin rate; λ_{+C} is the upward capacity transition rate; and λ_{-L} is the downward load transition rate.

The probability of the margin state is defined as the product of capacity state and the load state probabilities. It can be written in the following form:

$$P_k = P_C P_i$$

At the same time, a Reserve Margin is the percentage of additional installed capacity through a period that the annual peak demand is recorded. By defining a target generation margin, it is a deterministic criteria used to evaluate system reliability. To calculate the reserve margin, the following equation is used:

$$\text{Reserve Margin} = \frac{\text{Installed Capacity}\left(\text{MW}\right) - \text{Peak Load}\left(\text{MW}\right)}{\text{Peak Load}\left(\text{MW}\right)} \times 100\%$$

As an example (Figure 6.7), consider an electric generation system having 3 × 50MW units, where each unit has a failure rate (λ) of 0.02 f/year and a repair (μ) of 0.48 r/year as shown by the following state-space diagram:

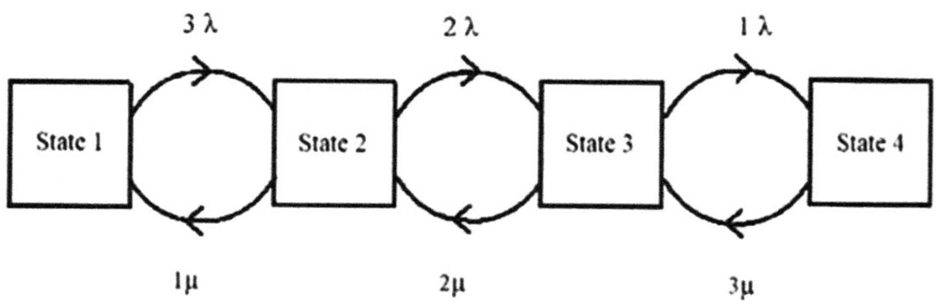

FIGURE 6.7 Four-state space diagram model.

Also, consider for the same system: the variation of a load over a 10 days period (D) scheduled as shown in Table 6.7.

And assume that the exposure factor (e) = 0.5 of a day. Find the capacity margin probability by combining both the generation and load models. The calculations must be on annual basis.

TABLE 6.7 Load Variation Over 10 Days Period

State (i)	Peak Load Level, L_i (MW)	No. of Occurrence, n_i
1	120	2
2	80	3
3	40	5

In this case, the following solution for the case is followed:

$$p = \frac{\mu}{\mu + \lambda}$$

$$q = \frac{\lambda}{\mu + \lambda}$$

Therefore,

$$p = 0.96$$
$$q = 0.04$$
$$e = 0.5$$

Then, calculating the individual probabilities that the electric power system is passing through using:

$$P_i = \frac{n!}{n!(n-r)!} p^r q^{(n-r)}$$

$$P_1 = \frac{3!}{3!(0)!} 0.96^3 0.04^0$$

$$P_1 = 0.884736$$

$$P_2 = \frac{3!}{2!(1)!} 0.96^2 0.04^1$$

$$P_2 = 0.110592$$

$$P_3 = \frac{3!}{1!(2)!} 0.96^1 0.04^2$$

$$P_3 = 0.004608$$

$$P_4 = \frac{3!}{0!(3)!} 0.96^0 0.04^3$$

$$P_4 = 0.000064$$

The required calculation for the generation model is summarized in Table 6.8, and the load model in Table 6.9.

TABLE 6.8 Generation Model Calculation

State	Capacity Out (MW)	Capacity IN (MW)	Individual Probability	Departure Rate		Frequency
				λ_{+n}	λ_{-C}	
1	0	150	0.884736	0	0.06	0.053084
2	50	100	0.110592	0.48	0.04	0.057507
3	100	50	0.004608	0.96	0.02	0.004516
4	150	0	0.000064	1.44	0	0.000092

TABLE 6.9 Load Model Calculation

State	Peak Load (MW), L_i	Number of Occurrence (n_i)	Probability $=\frac{n_i\, e}{D}$	Departure Rate	
				λ_{+L}	λ_{-L}
1	120	2	0.1	0	2
2	80	3	0.15	0	2
3	40	5	0.25	0	2
0	0	$\Sigma = 10$	$1 - e = 0.5$	$\frac{1}{1-e} = 2$	0

The load model is converted to annual basis. The conversion results are given in Table 6.10.

TABLE 6.10 Load Model Conversion to Annual Basis

Probability	Annual Basis conversion (Probability $\times \frac{10}{365}$
0.1	0.002739
0.15	0.004109
0.25	0.006849
0.5	0.013697

Since,

$$m_k = P_C - L_i$$

and

$$P_k = P_C . P_i$$

At the same time, departure rates written in the following form:

$$\lambda_{+m} = \lambda_{+C} + \lambda_{-L}$$

$$\lambda_{-m} = \lambda_{-C} + \lambda_{+L}$$

Therefore, the capacity margin and probability calculated as shown in Table 6.11.

TABLE 6.11 Capacity Margin and Probability

Capacity Margin (MW)	Individual Probability
150	0.012118
110	0.006059
100	0.001514
70	0.003635
60	0.000757
50	0.000063
30	0.002423
20	0.000454
10	0.000031
0	≈ 0.0
–10	-
–20	0.000303
–30	0.000018
–40	-

Table 6.12 merges the generation and the load shows the results. At the same time, the table shows the construction of margin-availability for the exact margin states. The states illustrated include all combinations of load and capacity.

6.3.3 CAPACITY RESERVE MODEL

In electrical power systems, a balance between supply and demand is expected. For an energy-producer, it refers to the capacity of energy-producer to generate more energy than the system normally requires. This means that a capacity reserve is available. The optimal spinning reserve is considered by the balance between the economy and the reliability of a power system (Li et al., 2019; Xie et al., 2018).

TABLE 6.12 Merging the Generation and the Load

Load	**1**	**2**	**3**	**0**
L_i (MW)	120	80	40	0
P_i	0.002739	0.004109	0.006849	0.013697
λ_{+L}	0	0	0	2
λ_{-L}	2	2	2	0
Generation				
n = 1, C =150 MW	m = 30	m = 70	m = 110	m = 150
P_1 = 0.884736	P= 0.002423	0.003635	0.006059	0.012118
λ_{+n}= 0	λ_+ = 2	2	2	0
λ_{-n}= 0.06	λ_- = 0.06	0.06	0.06	2.06
n = 2, C =100 MW	m = –20	m = 20	m = 60	m = 100
P_2 = 0.116592	P = 0.000303	0.000454	0.000757	0.001514
λ_{+n}= 0.48	λ_+ = 2.48	2.48	2.48	0.48
λ_{-n}= 0.04	λ_- = 0.04	0.04	0.04	2.04
n = 3, C = 50 MW	m = –70	m = –30	m = 10	m = 50
P_3 = 0.004608	P= 0.000012	0.000018	0.000031	0.000063
λ_{+n}= 0.96	λ_+ = 2.96	2.96	2.96	0.96
λ_{-n}= 0.02	λ_- = 0.02	0.02	0.02	2.02
n = 4, C = 0 MW	m = –120	m = –80	m = –40	m = 0
P_4 = 0.000064	P = -	-	-	-
λ_{+n}= 1.44	λ_+ = 3.44	3.44	3.44	1.44
λ_{-n}= 0	λ_- = 0	0	0	2

6.3.4 EXTENSION TO THE LOSS OF ENERGY CONCEPT

The energy might change from one form to another. Based on that, there will be a loss in that energy. Even though, when it is transformed from one form to another form, there will be some energy loss. This means that when energy is converted to a different form, some of that energy is turned into a highly disordered form of energy. This means that part of that energy will be lost as heat. The main objective of reducing the electricity losses is helping to achieve the goal of universal access to electricity services. Lowering electricity loose-jointed with the greater financial sustainability of utilities. Moreover, lowering the electricity losses is helping to contribute to reduce the air pollution emissions. This means that the electricity losses represent a costly problem in countries and region (Usman et al., 2018).

Two important considerations are known with regard to technical losses. The first technical loss is the availability of power current how it will vary. This means that how much is the current flow through the system. At the same time, the technical losses tend to go up as load increases. The second technical loss is the distance from the source.

6.3.5 PJM METHODS AND MODIFIED PJM

The term PJM is related to Pennsylvania-New Jersey-Maryland Interconnection in the USA (Mid-Atlantic region power pool). The concept remains as a basic method to evaluate unit commitment risk. At the same time, it is used to evaluate the operational reserve requirements. The basic PJM method is followed to evaluate the probability of the committed generation to satisfy of fail to satisfy the expected demand through a period. This means it is used to evaluate the probability to the committed generation. In the end, the method is simplifying the electric power system representation (Billinton and Allan, 1996; Cui et al., 2017; Resource Adequacy Analysis Subcommittee, 2018). The PJM, in its basic form, simplifies the system representation and determines the adequacy of installed capacity requirements using LOLE. In other words, any generating-unit is represented by the two-state model, where the possibility of repair during the lead-time is neglected. It has to take into consideration that the criteria of the PJM are to keep the load in a way that shall not exceed the available capacity. PJM's capacity market ensures long-term grid reliability. The PJM capacity market is called the Reliability Pricing Model. The PJM is applied as a test system to secure the appropriate amount of power supply resources needed to meet predicted energy demand in the future. In an outage, the probability of finding a two-state model (Figure 6.8) on the outage at a time t:

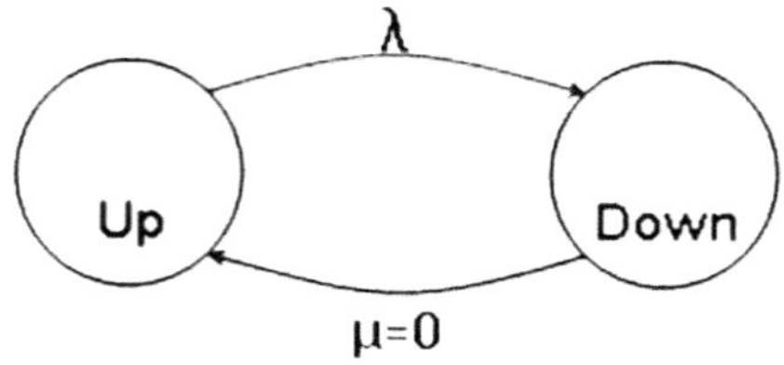

FIGURE 6.8 Non-repairable two-state model.

$$P_{down} = \frac{\lambda}{\lambda + \mu} - \frac{\lambda}{\lambda + \mu} e^{-(\lambda+\mu)t}$$

In case, the repair is neglected, the P_{down} is becoming:

$$P_{down} = 1 - e^{-\lambda t}$$

In cast that $\lambda t << 1$.

This case is applied for short lead-time to several hours. This is becoming:

$$P_{down} \approx \lambda t = \text{Outage Replacement Rate (ORR)}$$

This equation representing a generating-unit fails and not replaced by a new unit during the lead-time t. The ORR for each generating-unit is used instead of the FOR.

Example (6.8):

Consider a generating system consists of a number of units; 3 × 20MW, 2 × 30MW, and 3 × 50MW. Assuming that each generating-unit has a failure rate as shown in Table 6.13.

TABLE 6.13 Example (6.8) Data

Sub-System	Each Unit (MW)	Number of Unit(s)	Failure Rate (f/year)
A	20	2	3
B	30	3	3
C	50	2	4

Calculate the ORR for each case shown in Table 6.13 for lead-times 1, 2, 4 hours.

Solution:

Since the subsystems A and B having the same failure-rate equal 3 f/year, therefore:

$$ORR \approx \lambda\, t$$

$$ORR \approx \frac{3}{Year}(1\text{ hour})$$

$$ORR \approx \frac{3}{(365\text{ days} \times 24\text{ hours})}(1\text{ hour})$$

$$ORR \approx 0.000342$$

This is the ORR for each unit of sub-systems A and B. To calculate the ORR for sub-system C for the same period, which is 1 hour, the ORR becomes:

$$ORR \approx \lambda\, t$$

$$ORR \approx \frac{4}{Year}(1\text{ hour})$$

$$ORR \approx \frac{4}{(365\text{ days} \times 24\text{ hours})}(1\text{ hour})$$

$$ORR \approx 0.000457$$

Following the same procedure to find the ORR for the three sub-systems mentioned earlier. The results are summarized in Table 6.14.

TABLE 6.14 Results of Example (6.8)

Unit (MW)	Failure Rate (f/year)	ORR (for Lead-Time of Each Sub-System)		
		1 Hour	2 Hours	4 Hours
20	3	0.000342	0.000685	0.001370
30	3	0.000342	0.000685	0.001370
50	4	0.000457	0.000913	0.001826

Example (6.9):

Consider a generating power station consists of a number of units with their failure rates as stated in Table 6.15.

TABLE 6.15 Generating Power Station

Capacity Unit (MW)	Failure Rate (f/Yr)
30	2
20	4
50	4.5
40	5
60	5.5
55	6
45	6.5
35	7
70	8

Calculate the ORR for each case shown Table 6.16 for lead-times 1, 2, 3, 4 up to 8 hours. Then, sketch the variation of the ORR and the failure rates.

Solution:

Since,

$$ORR \approx \lambda\, t$$

Therefore, the results are summarized in Table 6.16. The results are represented by the following graph (Figure 6.9).

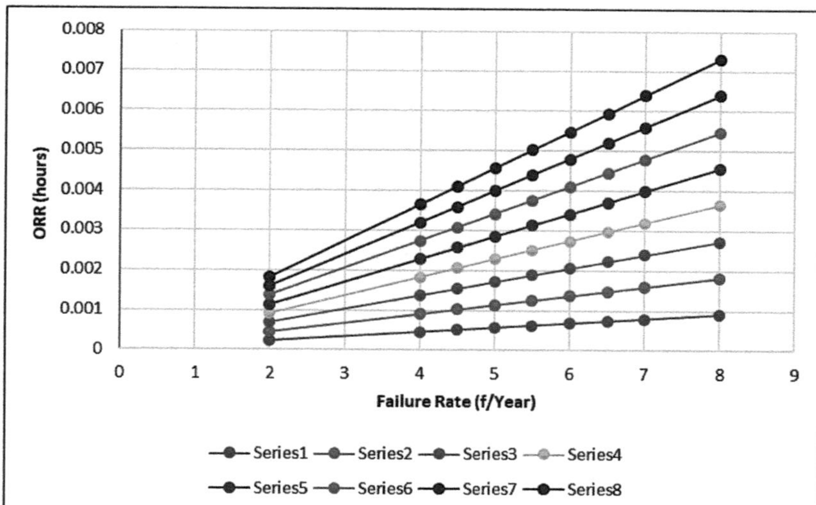

FIGURE 6.9 **(See color insert.)** The final results.

TABLE 6.16 Summary of the Results for Example (6.9)

Unit (MW)	Failure Rate (f/Yr)	ORR (hr)	ORR (hrs)	ORR (hrs)	ORR (hrs)	ORR (hrs)	ORR (hrs)	ORR (hrs)	ORR (hrs)
		1	2	3	4	5	6	7	8
30	2	0.000228	0.000456	0.000684	0.000913	0.001141	0.001369	0.001598	0.001826
20	4	0.000456	0.000913	0.001369	0.001826	0.002283	0.002739	0.003196	0.003652
50	4.5	0.000513	0.001027	0.001541	0.002054	0.002568	0.003082	0.003595	0.004109
40	5	0.000570	0.001141	0.001712	0.002283	0.002853	0.003424	0.003995	0.004566
60	5.5	0.000627	0.001255	0.001883	0.002511	0.003139	0.003767	0.004394	0.005022
55	6	0.000684	0.001369	0.002054	0.002739	0.003424	0.004109	0.004794	0.005479
45	6.5	0.000742	0.001484	0.002226	0.002968	0.003710	0.004452	0.005194	0.005936
35	7	0.000799	0.001598	0.002397	0.003196	0.003995	0.004794	0.005593	0.006392
70	8	0.000913	0.001826	0.002739	0.003652	0.004566	0.005479	0.006392	0.007305

Example (6.10):

Consider a generating power station consists of a number of units with their failure rates as stated in Table 6.17.

TABLE 6.17 Generating Power Station

Capacity Unit (MW)	Failure Rate (f/Yr)
30	0.2
20	0.3
25	0.35
40	0.4
35	0.45
50	0.5
55	0.55
60	0.6
45	0.7

Calculate the ORR for each case shown Table 6.17 for lead-times 1, 2, 3, 4 up to 8 hours. Then, sketch the variation of the ORR and the failure rates. Furthermore, sketch the histogram for the same variables.

Solution:

Since,

$$\mathrm{ORR} \approx \lambda\, \mathrm{t}$$

$$\mathrm{ORR} \approx \frac{4}{Month}(1\ \mathrm{hour})$$

$$\mathrm{ORR} \approx \frac{4}{(30\ \mathrm{days} \times 24\ \mathrm{hours})}(1\ \mathrm{hour})$$

$$\mathrm{ORR} \approx 0.005555556$$

Therefore, the results are summarized in Table 6.18, Figure 6.10, and Figure 6.11.

This case of the de-rated state which is called partial output state. This can be illustrated through a system presented as Figure 6.12, where this model can be reduced to Figure 6.13, which shows with a failure rates and no-repair rates. Figure 6.14 represents a model with no transition rates

TABLE 6.18 ORR Results

Failure Rate (f/ Month)	ORR (hr)	ORR (hrs)	ORR (hrs)	ORR (hrs)	ORR (hrs)	ORR (hrs)	ORR (hrs)	ORR (hrs)
	1	**2**	**3**	**4**	**5**	**6**	**7**	**8**
0.2	0.000278	0.000556	0.000833	0.001111	0.001389	0.001667	0.001944	0.002222
0.3	0.000417	0.000833	0.00125	0.001667	0.002083	0.0025	0.002917	0.003333
0.35	0.000486	0.000972	0.001458	0.001944	0.002431	0.002917	0.003403	0.003889
0.4	0.000556	0.001111	0.001667	0.002222	0.002778	0.003333	0.003889	0.004444
0.45	0.000625	0.00125	0.001875	0.0025	0.003125	0.00375	0.004375	0.005
0.5	0.000694	0.001389	0.002083	0.002778	0.003472	0.004167	0.004861	0.005556
0.55	0.000764	0.001528	0.002292	0.003056	0.003819	0.004583	0.005347	0.006111
0.6	0.000833	0.001667	0.0025	0.003333	0.004167	0.005	0.005833	0.006667
0.7	0.000972	0.001944	0.002917	0.003889	0.004861	0.005833	0.006806	0.007778

between the de-rated and down states. Figure 6.14 when considered and taking into consideration the assumption that $\lambda t \ll 1$, the following probabilities are obtained as:

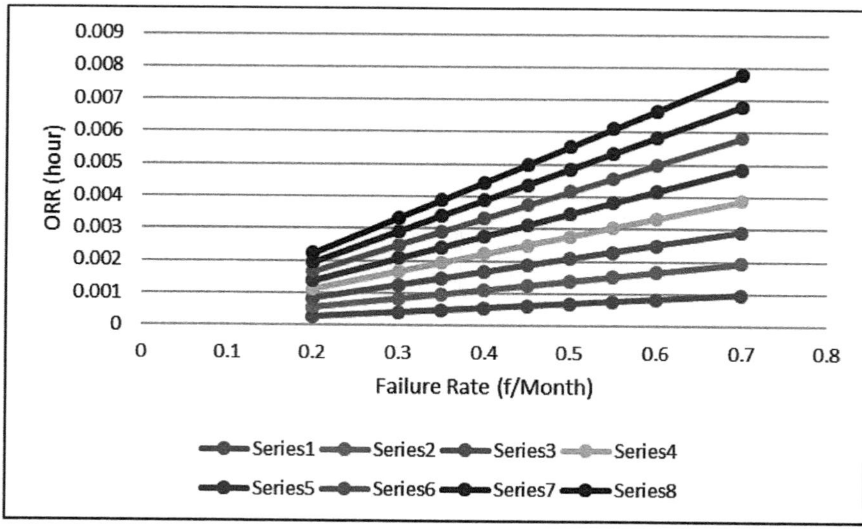

FIGURE 6.10 **(See color insert.)** ORR vs. failure rate relationship.

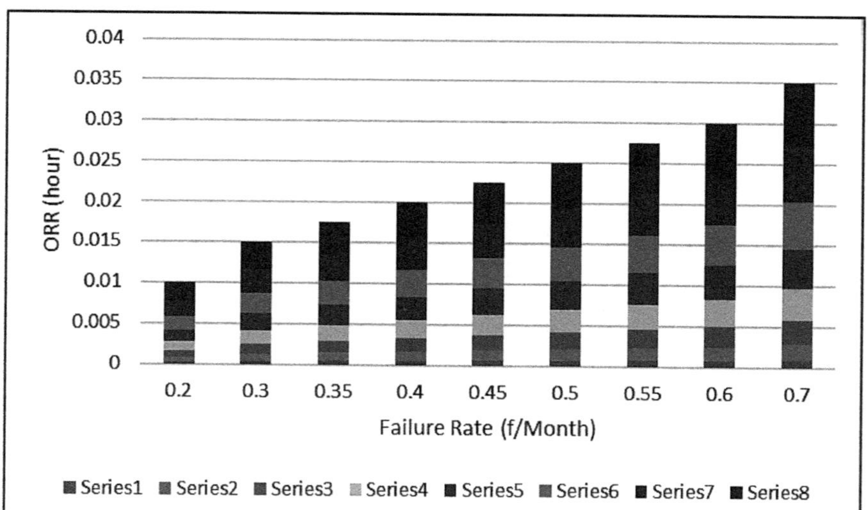

FIGURE 6.11 **(See color insert.)** ORR vs. failure rate histogram.

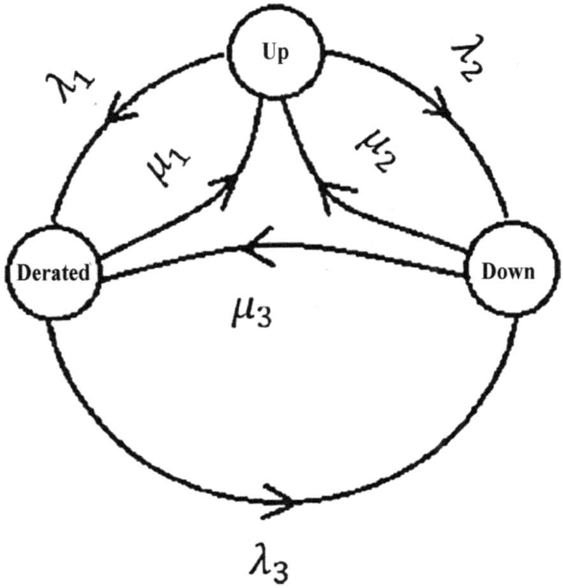

FIGURE 6.12 Three-state model of a generator.

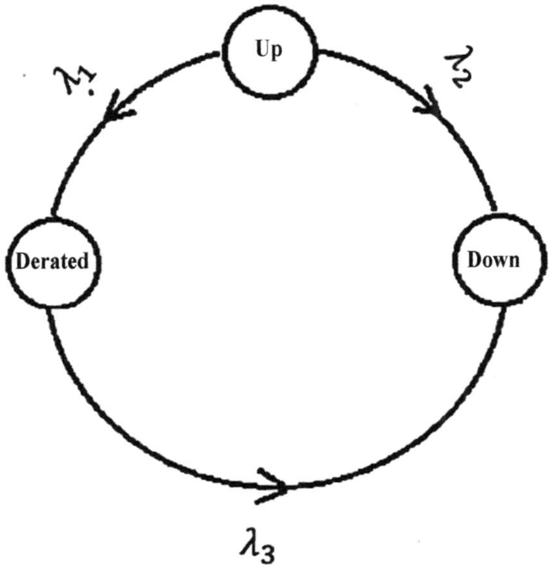

FIGURE 6.13 Three-state model of a generator after reduction.

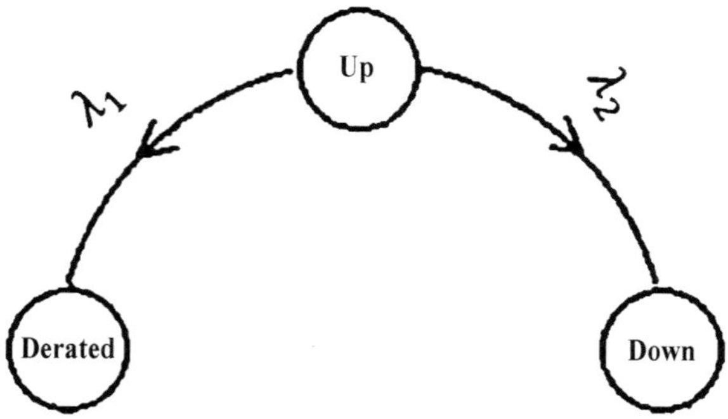

FIGURE 6.14 Three-state model of a generator with further reduction.

$$P(derated) = \lambda_1 t$$
$$P(down) = \lambda_2 t$$
$$P(Up) = 1 - (\lambda_1 + \lambda_2 t)$$

The modified PJM means that adding a new component(s) or replacing some by other(s) to satisfy the results or getting better results. The PJM can be extended, because of the assumption that the load can't be fixed. This means the load from year to year it is expected to increase and vary. Therefore, an estimated technique might be used for future expectations. This expectation needs to make a PJM modification. It is clear, in this case, to extend the system by standby generators or adding for future a new generators to avoid the risk.

6.3.6 RELIABILITY EVALUATION METHODS

To measure the reliability of a model (Gold Book, 2007), the analytical or digital methods are used. There are three methods used to measure the reliability system. These methods are the state space, network reduction, and cut-set methods. The state-space is considered as a general method but becoming cumbersome for large systems. The second method is the network reduction method is applicable for the series and parallel subsystems. The third method (cut-set method) becoming the most popular method. This method used for the transmission and distribution networks (Billinton and Allan, 1996).

6.3.7 DETERMINISTIC METHODS

In an electric power system, spinning reserve requirements can be determined by using deterministic and/or probabilistic methods (Billinton and Allan, 1996).

6.3.8 PLANNING RESERVE MARGIN (PRM)

Planning reserve margin (PRM) used in generation planning (Reimers et al., 2019). The objective of that is to determine an electric utility's resource. The PRM is used as a figure represents the value of probabilities of electric load occurrence (as an example) in a calculation for system reliability. Furthermore, it is useful in informing resource decisions between reliability indices studies. The peak load is calculated by following several factors. One factor is using the median annual peak load. Another factor is the generation resources, which are subject to forced and planned outages, which might be unavailable during some sometimes through the year when needed. As a third factor, when an interconnection network is available, the utilities hold operating reserves to keep the interconnection network reliable. These factors will keep the network under planning reserves conditions. Therefore, the PRM is well known as the percentage by which the total capacity of system resources exceeds the median peak load. To ensure that the supply of resources to be sufficient to meet load under different system conditions, the surplus capacity is very important.

The capacity reserve margin (CRM) is defined as:

$$\text{Capacity Reserve Margin}\left(\text{CRM}\right) = \frac{\left(\text{Dependable Resource Capacity} - \text{Expected Peak}\right)}{\text{Dependable Resource Capacity}}$$

A basic assumption is that enough resources exist to meet the expected load. The PRM is needed for two main reasons; these are the load and resources. The load for a short-term weather-related and Long-term load growth. Regarding the second reason, which is the resources, it depends on three main risks, which are the generation risk, the contract risk, and market risk.

6.3.9 CONTINGENCY RESERVE

The contingency reserve is an appropriation of surplus or retained earnings that may or may not be funded. It indicates a reservation against a specific or general contingency (Motalleb et al., 2016).

6.4 METHOD OF CALCULATION OF INDICES

There are a number of methods to find the indices (Billinton and Allan, 1996). One of which is the Monte Carlo method, which a reliable method and can find the needed indices. At the same time, the method needed in any study is based on the modeling of the different fault clearing technology algorithms considered at the analyses (Honrubia-Escribano et al., 2015).

6.4.1 LOSS OF LOAD PROBABILITY (LOLP, BASIC METHOD)

As explained in section (6.2.1), the LOLP was defined (Boroujeni et al., 2012; Rusin and Wojaczek, 2015; Shirvani et al., 2012). In the present section, it will be illustrated through an example re-calculated referred in reference published by (Boroujeni et al., 2012).

Example (6.11):

Consider a small power plant having two generating-units. Each unit is 100MW with FOR equal to 10%. Find the LOLP for the considered plant that operates to supply 130MW. Then, calculate the number of days of load loss.

Solution:

To find the LOLP, the cumulative probabilities are needed.

No. of Units IN	Capacity IN (MW)	Capacity OUT (MW)	Individual Probability	Cumulative Probability
2	200	0	0.81	1
1	100	100	0.18	0.19
0	0	200	0.01	0.01

When the power plant is operating at 130MW, it means that the LOLP is the Cumulative Probability of Capacity IN less than the 130MW. This means the LOLP is equal to 0.19 as shown in the above table.

Number of days of load loss, i.e., LOLE = 0.19 × 365 = 69.35 Days/year

Example (6.12):

Consider a power plant having three generators. Each generator is 50MW with FOR equal to 5%. Find the LOLP for the considered plant that operates to supply 120MW. Then, calculate the number of days of load loss.

Solution:

To find the LOLP, the cumulative probabilities are needed.

No. of Units IN	Capacity IN (MW)	Capacity OUT (MW)	Individual Probability	Cumulative Probability
3	150	0	0.857375	1
2	100	50	0.135375	0.142625
1	50	100	0.007125	0.00725
0	0	150	0.000125	0.000125

When the power plant is operating at 120MW, it means that the LOLP is the Cumulative Probability of Capacity IN less than the 120MW. This means the LOLP is equal to 0.142625, as shown in the above table.

Number of days of load loss, i.e., LOLE = 0.142625 × 365 = 52.1 days/year

Example (6.13):

In the power system planning (design, operation, and maintenance), reliability is considered as one of the factors that affect. At the same time, athe reliability is divided into adequacy and security. The adequacy relates to the available generation within the system to satisfy the ecustomers demand. Consider the system contains six generating units (Boroujeni et al., 2012) illustrated in Table 6.19. Calculate the probabilities for each company, then merge them to find the probabilities that the system might pass through. How to calculate the LOLP?

TABLE 6.19 System Contains Six Generating Units

Generation Company	Number of Generating-Units	Capacity of Each Generating-Unit (MW)	FOR
1	2	25	0.03
2	2	40	0.02
3	1	50	0.01
4	1	100	0.01

Company (1)	IN (MW)	OUT(MW)	Probability
2 × 25MW	50	0	0.9409
	25	25	0.0582
	0	50	0.0009
			Σ = 1
Company (2)	**IN (MW)**	**OUT (MW)**	**Probability**
2 × 40MW	80	0	0.9604
	40	40	0.0392
	0	80	0.0004
			Σ = 1
Company (3)	**IN (MW)**	**OUT (MW)**	**Probability**
1 × 50MW	50	0	0.99
	0	50	0.01
			Σ = 1
Company (4)	**IN (MW)**	**OUT (MW)**	**Probability**
1 × 100MW	100	0	0.99
	0	100	0.01
			Σ = 1

Connection(s), IN (MW)	IN (MW)	OUT (MW)	Probability
50 MW(1) + 80 MW(2) + 50 MW(3) + 100 MW(4)	280	0	0.885657917
25 MW + 80 MW + 50 MW + 100 MW	255	25	0.054782964
50 MW + 40 MW + 50 MW + 100 MW	240	40	0.036149303
(0 MW + 80 MW + 50 MW + 100 MW) & (50 MW + 80 MW + 0 MW + 100 MW)	230	50	0.009793199

Connection(s), IN (MW)	IN (MW)	OUT (MW)	Probability
25 MW + 40 MW + 50 MW + 100 MW	215	65	0.002236039
25 MW + 80 MW + 0 MW + 100 MW	205	75	0.000553363
50 MW + 0 MW + 50 MW + 100 MW	200	80	0.00036887
(0 MW + 40 MW + 50 MW + 100 MW) & (50 MW + 40 MW + 0 MW + 100 MW)	190	90	0.000399722
(0 MW + 80 MW + 0 MW + 100 MW) & (50 MW + 80 MW + 50 MW + 0 MW)	180	100	0.008954597
25 MW + 0 MW + 50 MW + 100 MW	175	105	2.28167E–05
25 MW + 40 MW + 0 MW + 100 MW	165	115	2.25863E–05
25 MW + 80 MW + 50 MW + 0 MW	155	125	0.000553363
(50 MW + 0 MW + 0 MW + 100 MW) & (0 MW + 0 MW + 50 MW + 100 MW)	150	130	4.0788E–06
(50 MW + 40 MW + 50 MW + 0 MW) & (0 MW + 40 MW + 0 MW + 100 MW)	140	140	0.000365494
(0 MW + 80 MW + 50 MW + 0 MW) & (50 MW + 80 MW + 0 MW + 0 MW)	130	150	9.89212E–05
25 MW + 0 MW + 0 MW + 100 MW	125	155	2.30472E–07
25 MW + 40 MW + 50 MW + 0 MW	115	165	2.25863E–05
25 MW + 80 MW + 0 MW + 0 MW	105	175	5.58953E–06
(50 MW + 0 MW + 50 MW + 0 MW) & (0 MW + 0 MW + 0 MW + 100 MW)	100	180	3.72953E–06
(50 MW + 40 MW + 0 MW + 0 MW) & (0 MW + 40 MW + 50 MW + 0 MW)	90	190	4.0376E–06
0 MW + 80 MW + 0 MW + 0 MW	80	200	8.6436E–08
25 MW + 0 MW + 50 MW + 0 MW	75	205	2.30472E–07
25 MW + 40 MW + 0 MW + 0 MW	65	215	2.28144E–07
(50 MW + 0 MW + 0 MW + 0 MW) + (0 MW + 0 MW + 50 MW + 0 MW)	50	230	4.12E–08
0 MW + 40 MW + 0 MW + 0 MW	40	240	3.528E–09
25 MW + 0 MW + 0 MW + 0 MW	25	255	2.328E–09
0 MW + 0 MW + 0 MW + 0 MW	0	280	3.6E–11

The LOLP is calculated as:

$$LOLP = \sum_{j=1}^{n} \frac{P_j t_j}{100}$$

where: P_j is the probability of capacity outage; t_j is the percentage of time when the load exceeds C_j; and C_j is the remaining generation capacity.

Appendix (C) illustrates the steps of calculating the LOLP.

6.5 FREQUENCY AND DURATION METHOD

An extension of LOLE is the frequency and duration (F and D) criterion (Billinton and Allan, 1992). The LOLE identifying the expected frequencies of encountering deficiencies and their expected durations. If a four-state diagram is considered and represented in Figure 6.15, where this diagram is known as a Markov state diagram model.

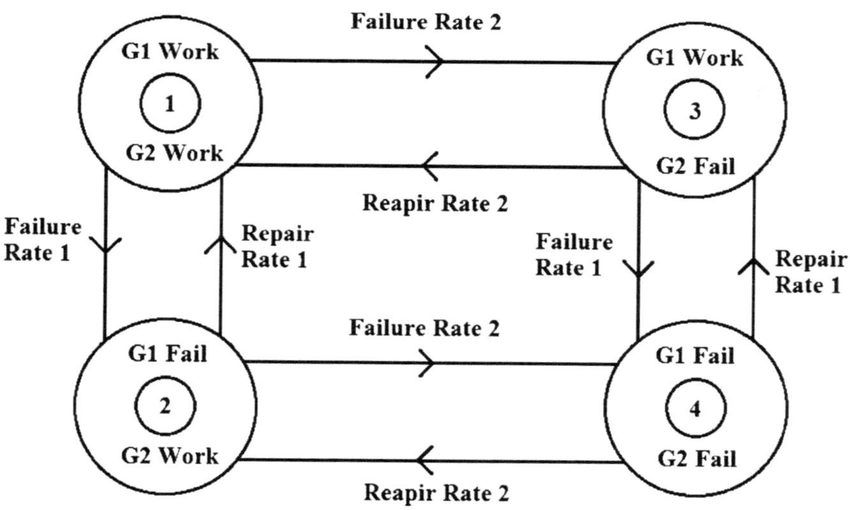

FIGURE 6.15 A two-generator state-space diagram model.

Figure 6.15 represents the four-state model, where the transition rates are defined as:

$$Failure\ Rate\ 1 = \lambda_1$$
$$Failure\ Rate\ 2 = \lambda_2$$
$$Repair\ Rate\ 1 = \mu_1$$
$$Repair\ Rate\ 2 = \mu_2$$

The steady-state probabilities were calculated in section 5.3.1 for the two-state model. The results of that clear that:

Any Generator in the working-state has a probability equal to:

$$\text{Working State Probability} = \frac{\mu}{\mu+\lambda}$$

$$\text{Failing State Probability} = \frac{\lambda}{\mu+\lambda}$$

Therefore, the State-probabilities for the first generator (G1) are:

$$\text{Working State Probability}(G1) = \frac{\mu_1}{\mu_1+\lambda_1}$$

$$\text{Failing State Probability}(G1) = \frac{\lambda_1}{\mu_1+\lambda_1}$$

and the State-probabilities for the first generator (G2) are:

$$\text{Working State Probability}(G2) = \frac{\mu_2}{\mu_2+\lambda_2}$$

$$\text{Failing State Probability}(G2) = \frac{\lambda_2}{\mu_2+\lambda_2}$$

At this stage, the probabilities of the four-state model (Figure 6.15) are calculated as:

$$\text{State}(1)\ \text{Probability} = \frac{\mu_1}{\mu_1+\lambda_1}\frac{\mu_2}{\mu_2+\lambda_2}$$

$$\text{State}(2)\ \text{Probability} = \frac{\lambda_1}{\mu_1+\lambda_1}\frac{\mu_2}{\mu_2+\lambda_2}$$

$$\text{State}(3)\ \text{Probability} = \frac{\mu_1}{\mu_1+\lambda_1}\frac{\lambda_2}{\mu_2+\lambda_2}$$

$$\text{State}(4)\ \text{Probability} = \frac{\lambda_1}{\mu_1+\lambda_1}\frac{\lambda_2}{\mu_2+\lambda_2}$$

The frequency of each state is calculated as follows:

$$f_1 = P(\text{State } 1)\ x\ (\lambda_1 + \lambda_2)$$

$$f_2 = P(State\ 2)\ x\ (\mu_1 + \lambda_2)$$
$$f_3 = P(State\ 3)\ x\ (\lambda_1 + \mu_2)$$
$$f_4 = P(State\ 4)\ x\ (\mu_1 + \mu_2)$$

The duration of each state is calculated as follows:

$$Duration(through\ State1) = \frac{State(1)\ Probability}{f_1}$$

$$Duration(through\ State2) = \frac{State(2)\ Probability}{f_2}$$

$$Duration(through\ State3) = \frac{State(3)\ Probability}{f_3}$$

$$Duration(through\ State4) = \frac{State(4)\ Probability}{f_4}$$

This means the duration in the Up-State becomes:

$$Duration(through\ Up\ State) = \frac{(Up - State).\ Probability}{f\,(Up - State)}$$

And the duration in the Down-State is:

$$Duration(through\ Down\ State) = \frac{(Down - State).\ Probability}{f\,(Down - State)}$$

6.6 MINIMAL CUT SET METHOD

The cut-set method is applied to a system that has components not connected in a simple form. It means they are not connected in a simple series/parallel form. As an example, the bridge circuit is considered as one form of that type of system. There are two main reasons to use the cut-set technique; these reasons are the simplicity of programming this type of logic network, and the second reason is the cut-set is related directly to the system modes (i.e., failure mode). For this reason, a suitable technique is required for application to solve this type of network. The solution needs to include the conditional probability method, cut, and tie set analysis and logic diagram. The main difference between methods is the logic. The system needs to be reduced to a number of sub-systems following the conditional probability method to reach a suitable solution. Otherwise, failure component(s) might cause a failure of system (Almuhaini and Al-sakkaf, 2017; Billinton and Allan, 1992; Kumar et al., 2017; Rebaiaia and Ait-Kadi, 2013; Zhao et al., 2018).

Example (6.14):

Consider a bridge circuit that has five components (X_1, X_2, Y_1, Y_2, and Z) connected, as shown in Figure 6.16a. What is the success needs to keep the system in a good condition? And find the overall reliability of the system. Then, assume that the components are identical. Calculate the overall reliability of the system when the reliability of each component is 0.95.

Solution:

Condition (1): If the component (Z) is good and replaced by a short circuit (allowing the Z to response, i.e., used to give a chance for response passing) as represented by Figure 6.16b. Therefore, the best connections to keep the system under operation are representing the reliability of the system as:

$$R_{system} = R_{X1X2}\, R_{Y1Y2}$$

$$R_{system} = (1 - Q_{X1}\, Q_{X2})\,(1 - Q_{Y1}\, Q_{Y2})$$

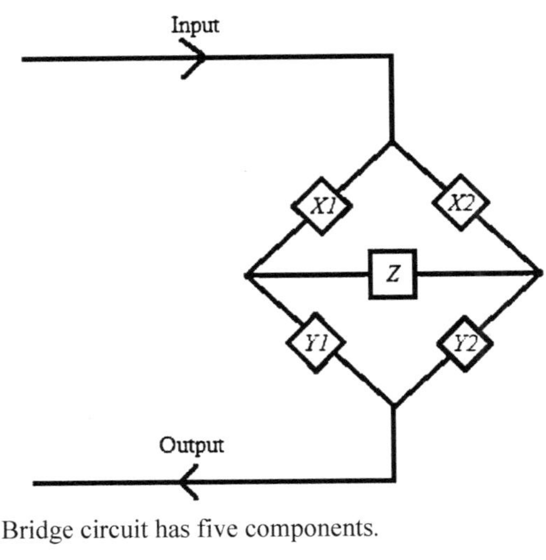

FIGURE 6.16a Bridge circuit has five components.

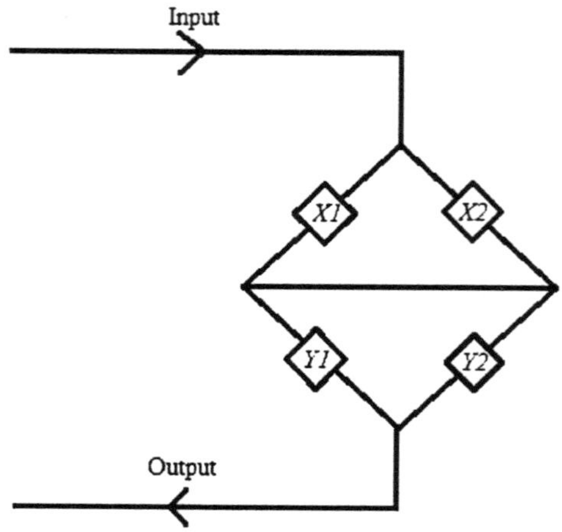

FIGURE 6.16b Bridge circuit has five components, Z is good (short circuit)

Condition (2): If the component (Z) is bad and replaced by the open circuit (i.e., the component Z does not exist). Therefore, the best connections to keep the system under operation are representing the reliability of the system as:

$$R_{system} = (Q_{X1Y1})\,(Q_{X2Y2})$$

$$R_{system} = (1 - R_{X1}\,R_{Y2})\,(1 - R_{Y1}\,R_{X2})$$

∴ The overall Reliability of the system (Figure 6.17) is:

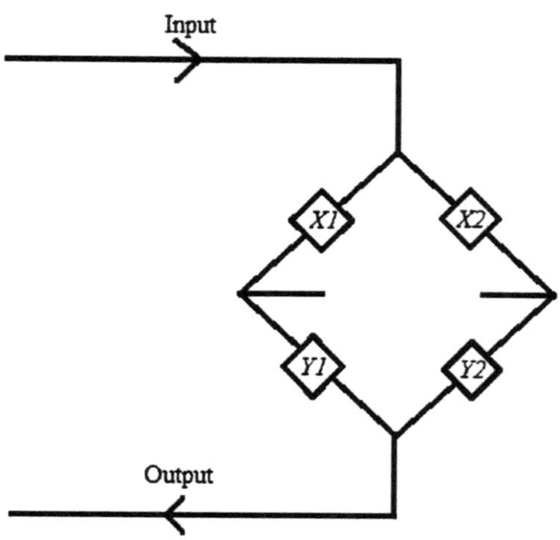

FIGURE 6.17 Bridge circuit has five components, Z is bad (open circuit).

$$R_{overall\ system} = (1-Q_{X1}Q_{X2})(1-Q_{Y1}Q_{Y2})R_Z + [1-(1-R_{X1}R_{Y2})(1-R_{Y1}R_{X2})]Q_Z$$

$$R_{overall\ system} = R_Z - Q_{X1}Q_{X2}R_Z - Q_{Y1}Q_{Y2}R_Z + Q_{X1}Q_{X2}Q_{Y1}Q_{Y2}R_Z$$
$$+ Q_Z - Q_Z(1-R_{X1}R_{Y2})(1-R_{Y1}R_{X2})$$

$$R_{overall\ system} = R_Z - (1-R_{X1})(1-R_{X2})R_Z - (1-R_{Y1})(1-R_{Y2})R_Z + (1-R_{X1})$$
$$(1-R_{X2})(1-R_{Y1})(1-R_{Y2})R_Z + 1 - R_Z - (1-R_Z)(1-R_{X1}R_{Y2})$$
$$(1-R_{Y1}R_{X2})$$

$$R_{overall\ system} = R_{X1}R_{Y1} + R_{X2}R_{Y2} + R_{X1}R_{Y2}R_Z + R_{X2}R_{Y1}R_Z - R_{X1}R_{X2}R_{Y1}R_{Y2}$$
$$- R_{X1}R_{Y1}R_{Y2}R_Z - R_{X1}R_{X2}R_{Y1}R_Z - R_{X2}R_{Y1}R_{Y2}R_Z - R_{X1}R_{X2}R_{Y2}R_Z$$
$$+ 2R_{X1}R_{X2}R_{Y1}R_{Y2}R_Z$$

This is the overall system reliability.

When the components are identical, and the reliability of each component is 0.95. Therefore, the overall reliability is:

$$R_{overall\ system} = R^2 + R^2 + R^3 + R^3 - R^4 - R^4 - R^4 - R^4 - R^4 + 2R^5$$

$$R_{overall\ system} = 2R^2 + 2R^3 - 5R^4 + 2R^5$$

In case that the reliability of each component is 0.95. Therefore, the overall reliability of the bridge circuit is calculated:

$$R_{overall\ system} = 2(0.95)^2 + 2(0.95)^3 - 5(0.95)^4 + 2(0.95)^5$$

$$R_{overall\ system} = 0.994780625$$

In case that the reliability of each component varies between 0.05 and 0.95 with a step of 0.5. Therefore, the overall reliability becoming as shown in Table 6.20 and drawn as illustrated in both Figures 6.18 and 6.19.

Example (6.15):

Consider a bridge network that has five transmission lines (T.L1, T.L2, T.L3, T.L4, and T.L5). The bridge network fed by a generator connected, as shown in Figure 6.20, and ended by a Load. What is the success needs to keep the system in a good condition? Then, assume that the transmission lines are identical. Calculate the overall reliability of the system when the reliability of each transmission line is 0.95 and the reliability of Generator is 0.91.

TABLE 6.20 Bridge Circuit Results

Each Component Reliability	Overall System Reliability
0.95	0.994780625
0.9	0.97848
0.85	0.950629375
0.8	0.91136
0.75	0.861328125
0.7	0.80164
0.65	0.733776875
0.6	0.65952
0.55	0.580875625
0.5	0.5
0.45	0.419124375
0.4	0.34048
0.35	0.266223125
0.3	0.19836
0.25	0.138671875
0.2	0.08864
0.15	0.049370625
0.1	0.02152
0.05	0.005219375

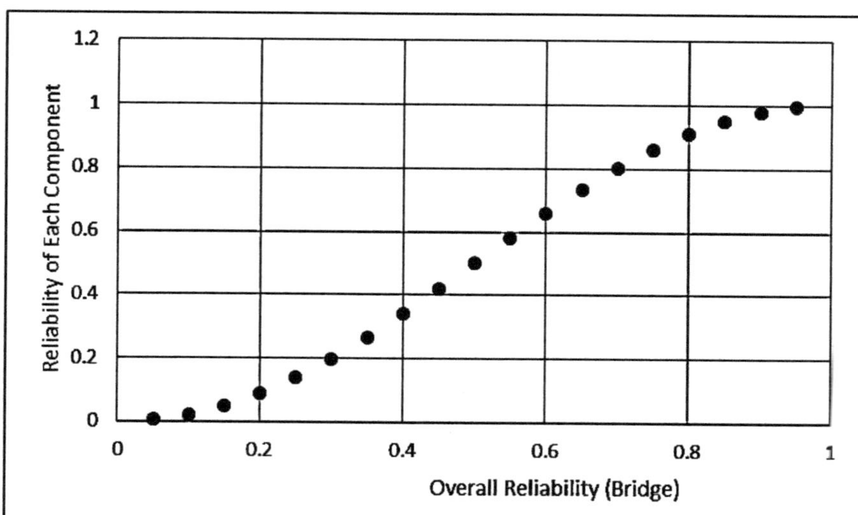

FIGURE 6.18 Bridge circuit results.

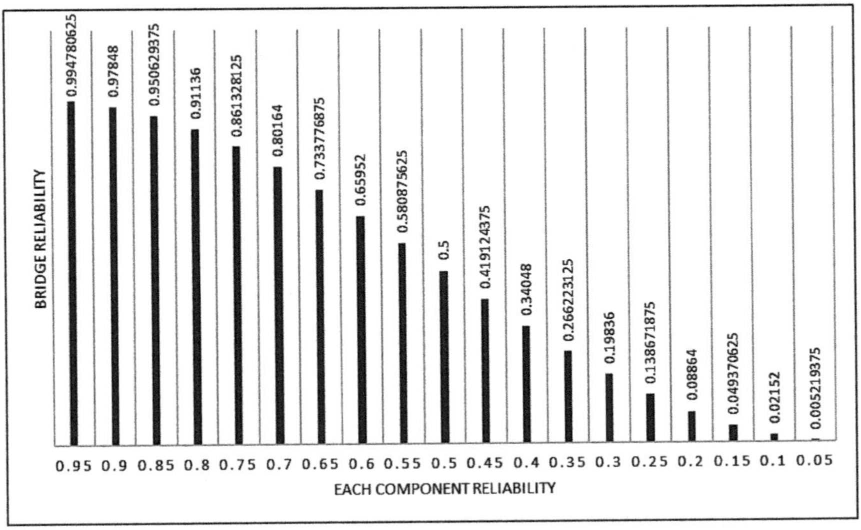

FIGURE 6.19 Bridge circuit results histogram.

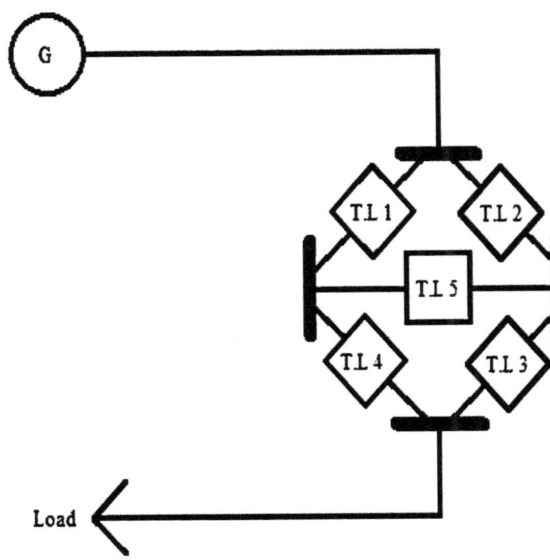

FIGURE 6.20 Bridge network.

Solution:

Figure 6.20 is converted to a tie-set diagram as shown in Figure 6.21.

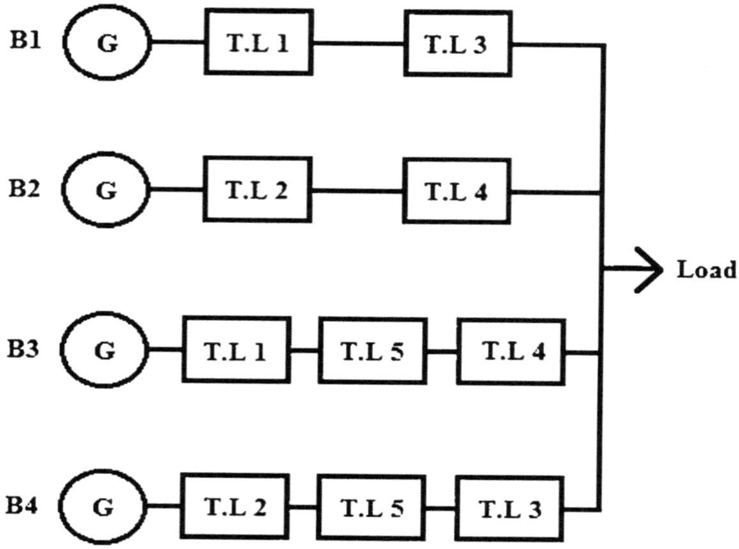

FIGURE 6.21 Tie-set diagram.

The present example is similar to the previous example (bridge circuit), except that the generator and the load are added to the bridge. Therefore, the bridge network has the overall reliability of the system:

$$R_{overall\ bridge} = R(B1) \cup R(B1) \cup R(B3) \cup R(B4)$$

This means that the overall bridge reliability of the bridge is equal to the probabilities of the branches.

$$\begin{aligned} R_{overall\ bridge} &= R(B1) + R(B2) + R(B3) + R(B4) \\ &- \{R(B1 \cap B2) + R(B1 \cap B3) + R(B1 \cap B4) - R(B2 \cap B3) \\ &- R(B2 \cap B4) - R(B3 \cap B4)\} + R(B1 \cap B2 \cap B3) + R(B1 \cap B2 \cap B4) \\ &+ R(B1 \cap B3 \cap B4) + R(B2 \cap B3 \cap B4) - R(B1 \cap B2 \cap B3 \cap B4) \end{aligned}$$

$$R(B1) = R(G)\ R(T.L1)\ R(T.L3)$$

$$R(B2) = R(G)\ R(T.L2)\ R(T.L4)$$

$$R(B3) = R(G)\ R(T.L1)\ R(T.L5)\ R(T.L4)$$

$$R(B4) = R(G)\ R(T.L2)\ R(T.L5)\ R(T.L3)$$

$$R(B1 \cap B2) = R(G)\,R(B1)\,R(B2) = R(G)\,R(T.L1)\,R(T.L2)\,R(T.L3)\,R(T.L4)$$

$$R(B1 \cap B3) = R(G)\,R(B1)\,R(B3) = R(G)\,R(T.L1)\,R(T.L3)\,R(T.L4)\,R(T.L5)$$

$$R(B1 \cap B4) = R(G)\,R(B1)\,R(B4) = R(G)\,R(T.L1)\,R(T.L2)\,R(T.L3)\,R(T.L5)$$

$$R(B2 \cap B3) = R(G)\,R(B2)\,R(B3) = R(G)\,R(T.L1)\,R(T.L2)\,R(T.L4)\,R(T.L5)$$

$$R(B2 \cap B4) = R(G)\,R(B2)\,R(B4) = R(G)\,R(T.L2)\,R(T.L3)\,R(T.L4)\,R(T.L5)$$

$$R(B3 \cap B4) = R(G)\,R(B3)\,R(B4) = R(G)\,R(T.L1)\,R(T.L2)\,R(T.L3)\,R(T.L4)\,R(T.L5)$$

$$\begin{aligned} R(B1 \cap B2 \cap B3) &= R(B1 \cap B2 \cap B4) = R(B1 \cap B3 \cap B4) \\ &= R(B2 \cap B3 \cap B4) = R(B1 \cap B2 \cap B3 \cap B4) \\ &= R(G)\,R(T.L1)\,R(T.L2)\,R(T.L3)\,R(T.L4)\,R(T.L5) \end{aligned}$$

Therefore, the overall reliability of the bridge network becoming 0.905250369 and the un-reliability is 0.094749631. The Up-Time of the network is 7929.99323 hours in a year, where the Down-Time of the network is 830.0067698 hours in a year.

In case the network has different values of the generator-reliability and different values of each transmission lines reliability. Therefore, the results are recorded in Table 6.21.

TABLE 6.21 Bridge Network Results

G	TL	$R_{overall}$	$Q_{overall}$	Up Time (Hour)	Dn Time (Hour)
0.92	0.95	0.915198175	0.084801825	8017.136	742.864
0.95	0.99	0.949808147	0.050191853	8320.319	439.6806
0.94	0.98	0.939233706	0.060766294	8227.687	532.3127
0.93	0.97	0.928279501	0.071720499	8131.728	628.2716
0.92	0.96	0.916949828	0.083050172	8032.48	727.5195
0.91	0.95	0.905250369	0.094749631	7929.993	830.0068
0.9	0.94	0.89318812	0.10681188	7824.328	935.6721
0.89	0.93	0.880771313	0.119228687	7715.557	1044.443
0.88	0.92	0.868009337	0.131990663	7603.762	1156.238
0.87	0.91	0.854912669	0.145087331	7489.035	1270.965
0.86	0.9	0.8414928	0.1585072	7371.477	1388.523
0.85	0.89	0.827762164	0.172237836	7251.197	1508.803

TABLE 6.21 *(Continued)*

G	TL	$R_{overall}$	$Q_{overall}$	Up Time (Hour)	Dn Time (Hour)
0.84	0.88	0.813734068	0.186265932	7128.31	1631.69
0.83	0.87	0.799422627	0.200577373	7002.942	1757.058
0.82	0.86	0.784842693	0.215157307	6875.222	1884.778
0.81	0.85	0.770009794	0.229990206	6745.286	2014.714
0.8	0.84	0.754940068	0.245059932	6613.275	2146.725
0.79	0.83	0.739650202	0.260349798	6479.336	2280.664
0.78	0.82	0.724157371	0.275842629	6343.619	2416.381
0.77	0.81	0.708479179	0.291520821	6206.278	2553.722
0.76	0.8	0.6926336	0.3073664	6067.47	2692.53
0.75	0.79	0.676638922	0.323361078	5927.357	2832.643
0.74	0.78	0.660513694	0.339486306	5786.1	2973.9
0.73	0.77	0.64427667	0.35572333	5643.864	3116.136
0.72	0.76	0.627946758	0.372053242	5500.814	3259.186
0.71	0.75	0.611542969	0.388457031	5357.116	3402.884
0.7	0.74	0.595084367	0.404915633	5212.939	3547.061

In case that the reliability of the generator is fixed, where the reliability of the identical transmission lines vary as shown in Table 6.22.

TABLE 6.22 Reliability of the Identical Transmission Lines

G	TL	$R_{overall}$	$Q_{overall}$	Up Time (Hour)	Dn Time (Hour)
0.95	0.95	0.945041594	0.054958406	8278.564	481.4356
0.95	0.99	0.949808147	0.050191853	8320.319	439.6806
0.95	0.98	0.949225554	0.050774446	8315.216	444.7841
0.95	0.97	0.948242501	0.051757499	8306.604	453.3957
0.95	0.96	0.946850365	0.053149635	8294.409	465.5908
0.95	0.95	0.945041594	0.054958406	8278.564	481.4356
0.95	0.94	0.942809683	0.057190317	8259.013	500.9872
0.95	0.93	0.940149154	0.059850846	8235.707	524.2934
0.95	0.92	0.937055534	0.062944466	8208.606	551.3935
0.95	0.91	0.933525328	0.066474672	8177.682	582.3181
0.95	0.9	0.929556	0.070444	8142.911	617.0894
0.95	0.89	0.925145948	0.074854052	8104.279	655.7215
0.95	0.88	0.920294482	0.079705518	8061.78	698.2203
0.95	0.87	0.915001802	0.084998198	8015.416	744.5842
0.95	0.86	0.909268973	0.090731027	7965.196	794.8038
0.95	0.85	0.903097906	0.096902094	7911.138	848.8623
0.95	0.84	0.896491331	0.103508669	7853.264	906.7359

TABLE 6.21 *(Continued)*

G	TL	$R_{overall}$	$Q_{overall}$	Up Time (Hour)	Dn Time (Hour)
0.95	0.83	0.889452775	0.110547225	7791.606	968.3937
0.95	0.82	0.881986542	0.118013458	7726.202	1033.798
0.95	0.81	0.874097689	0.125902311	7657.096	1102.904
0.95	0.8	0.865792	0.134208	7584.338	1175.662
0.95	0.79	0.857075968	0.142924032	7507.985	1252.015
0.95	0.78	0.84795677	0.15204323	7428.101	1331.899
0.95	0.77	0.838442242	0.161557758	7344.754	1415.246
0.95	0.76	0.828540861	0.171459139	7258.018	1501.982
0.95	0.75	0.818261719	0.181738281	7167.973	1592.027
0.95	0.74	0.807614499	0.192385501	7074.703	1685.297

Both curves (Figures 6.22 and 6.23) having the same shape, but with a different purpose.

Applying the curve-fitting for the four cases, with a fixed value of the generator-reliability equal to 0.95, the curves are obtained with their formulas:

a. Overall-reliability versus transmission lines reliabilities, represented by Figure 6.24 with the formula:

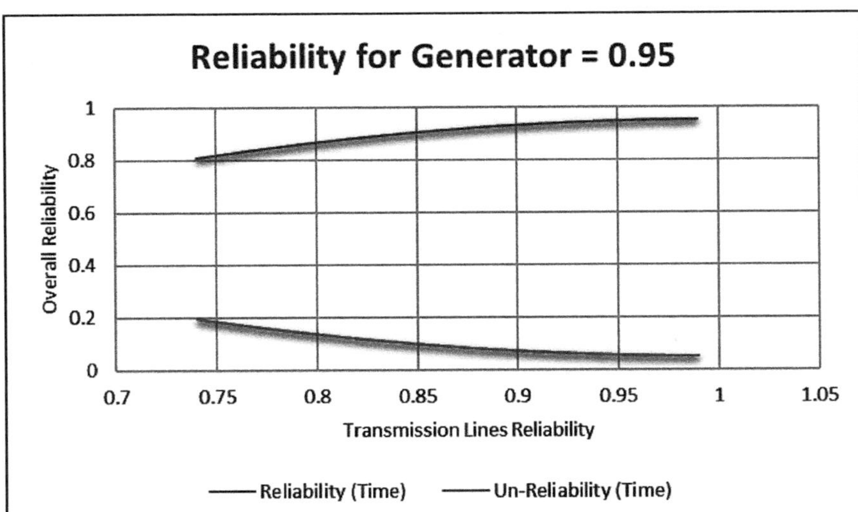

FIGURE 6.22 **(See color insert.)** Overall reliability vs. T.Ls reliability.

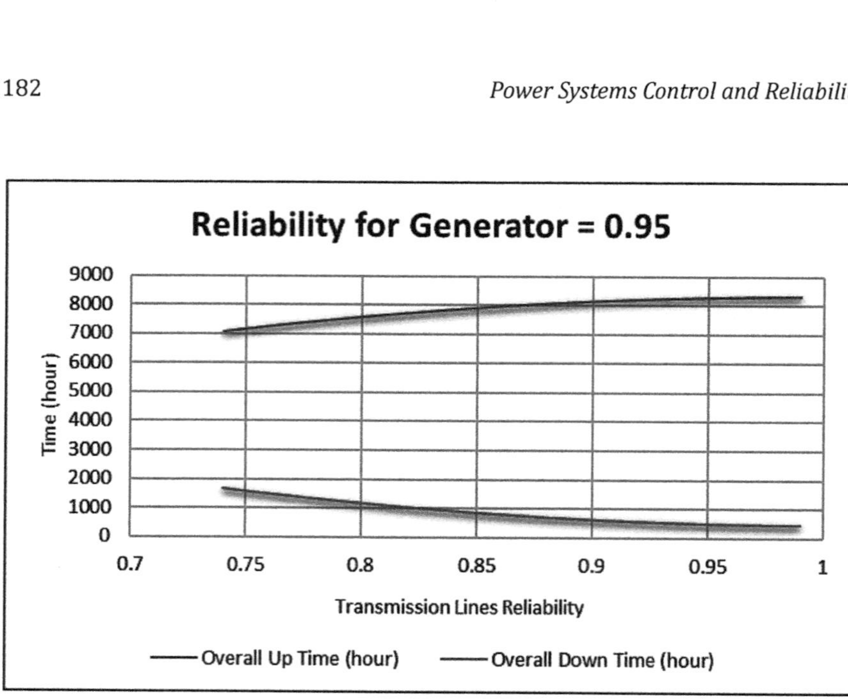

FIGURE 6.23 **(See color insert.)** Time vs. T.Ls reliability.

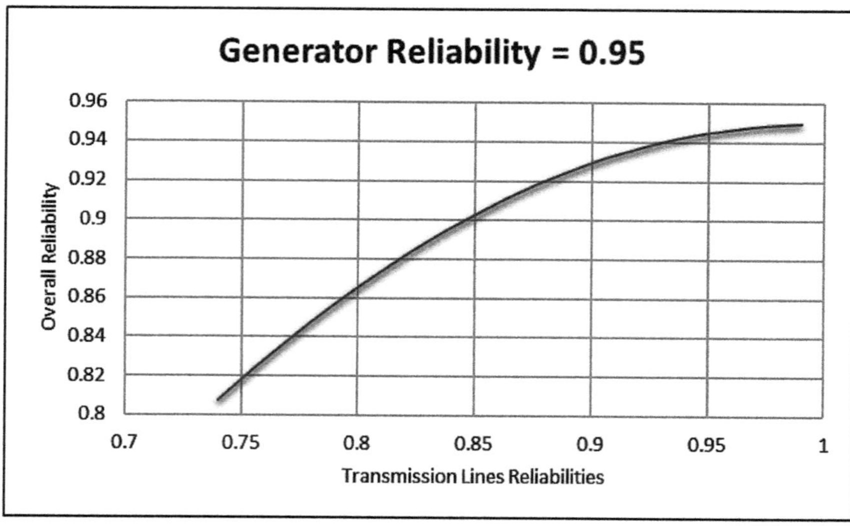

FIGURE 6.24 **(See color insert.)** Overall reliability vs. transmission lines reliabilities.

The formula obtained by the curve-fitting and represented as a fourth order polynomial as:

$$y = 0.8966963 - 5.236757x + 14.10175x^2 - 12.27874x^3 + 3.467044x^4$$

$a = 0.8966963 \pm 0.01654$
$b = -5{,}236757 \pm 0.07718$
$c = 14.10175 \pm 0.01346$
$d = -12.27874 \pm 0.104$
$e = 3.467044 \pm 0.03006$

b. Overall un-reliability versus transmission lines reliabilities, represented by Figure 6.25 with the formula:

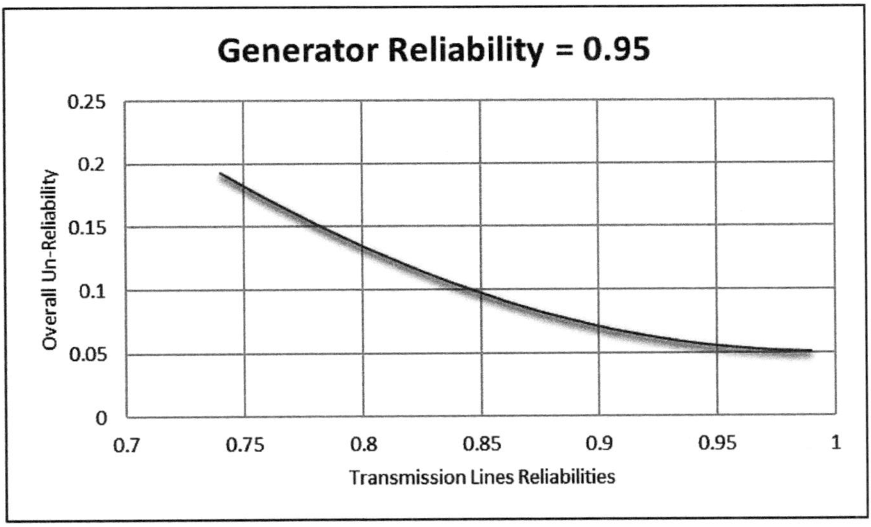

FIGURE 6.25 (See color insert.) Overall un-reliability vs. transmission lines reliabilities.

The formula obtained by the curve-fitting and represented as a fourth order polynomial as:

$$y = 0.1033037 + 5.236757x - 14.10175x^2 + 12.27874x^3 - 3.467044x^4$$

$a = 0.1033037 \pm 0.01654$
$b = 5.236757 \pm 0.07718$
$c = -14{,}101175 \pm 0.01346$
$d = 12.27874 \pm 0.104$
$e = -3.467044 \pm 0.03006$

c. Overall up-time versus transmission lines reliabilities, represented by Figure 6.26 with the formula:

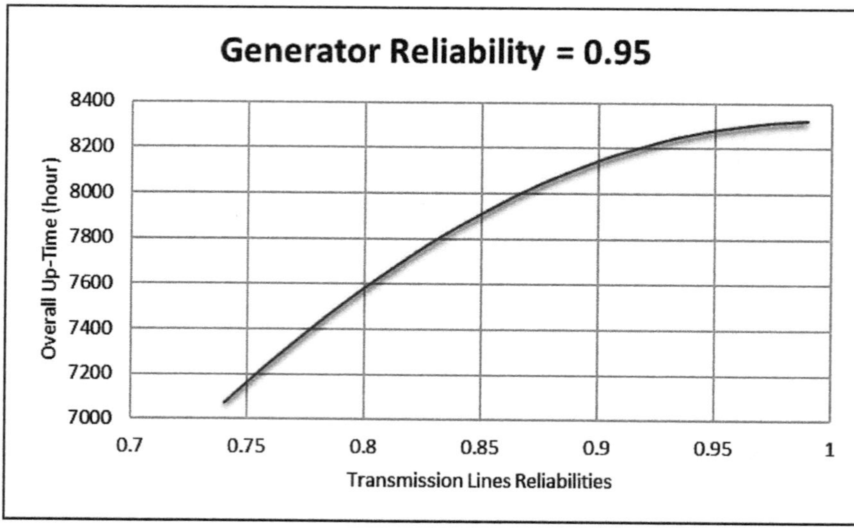

FIGURE 6.26 **(See color insert.)** Overall up-time (hour) vs. transmission lines reliabilities.

The formula obtained by the curve-fitting and represented as a fourth order polynomial as:

$$y = 7855.051 - 45873.95x + 123531.2x^2 - 107561.8x^3 + 30371.29x^4$$

$a = 7855.051 \pm 144.9$
$b = -45873.95 \pm 676.1$
$c = 123531.2 \pm 1179$
$d = -107561.8 \pm 911.3$
$e = 30371.29 \pm 263.3$

d. Overall Down-Time versus Transmission Lines Reliabilities, represented by Figure 6.27 with the formula:

The formula obtained by the Curve-Fitting and represented as a Fourth Order Polynomial as:

$$y = 904.9481 + 45873.95x - 123531.2x^2 + 107561.8x^3 - 30371.29x^4$$

$a = 904.9481 \pm 144.9$
$b = 45873.95 \pm 676.1$
$c = -123531.2 \pm 1179$
$d = 107561.8 \pm 911.3$
$e = -30371.29 \pm 263.3$

These four cases are helping the researcher for more study of such a case of bridge-networks or any type of networks that might need further study.

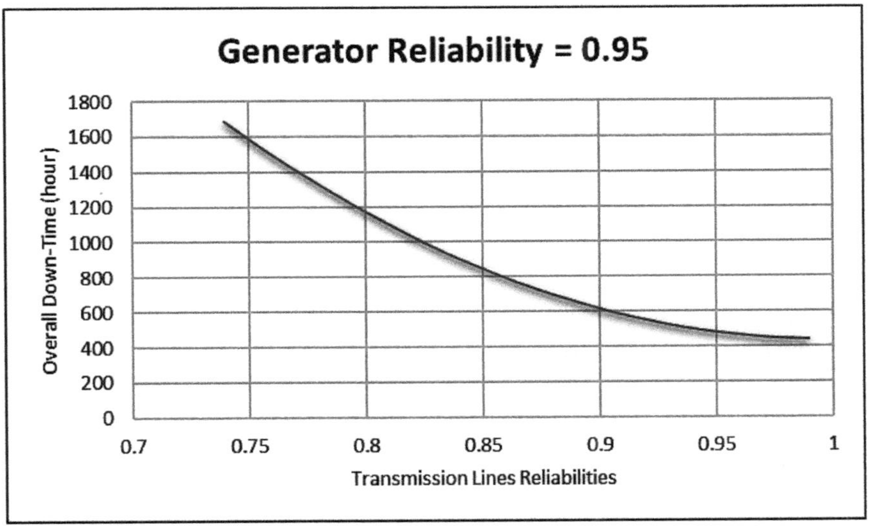

FIGURE 6.27 (See color insert.) Overall down-time (hour) vs. transmission lines reliabilities.

KEYWORDS

- **expected power not supplied**
- **expected un-served energy**
- **loss of energy probability**
- **loss of load events**
- **loss of load hours**
- **loss of load probability**

REFERENCES

Akhavein, A., & Porkar, B. A composite generation and transmission reliability test system for research purposes. *Renewable and Sustainable Energy Reviews*, **2017**, *75*, 331–337.

Almuhaini, M., & Al-Sakkaf, A. Markovian model for reliability assessment of microgrids considering load transfer. *Turkish Journal of Electrical Engineering and Computer Sciences*, **2017**, *25*, 4657–4672.

Al-Shaalan, A. M. Reliability evaluation in generation expansion planning based on the expected energy not served. *Journal of King Saud University – Engineering Sciences*, **2012**, *24*, 11–18.

Azizah, I. D., Abdullah, A. G., Purnama, W., Nandiyanto, A., & Shafii, M. A. Loss of load probability calculation for west java power system with nuclear power plant scenario. *1st Annual Applied Science and Engineering Conference, IOP Conf. Series: Materials Science and Engineering 180.* **2017**, 012079. doi: 10.1088/1757–899X/180/1/012079.

Beigvand, S. D., Abdi, H., & Scala, M. L. A general model for energy hub economic dispatch. *Applied Energy*, **2017**, *190*, 1090–1111.

Billinton, R., & Allan, R. N. *Reliability Evaluation of Engineering Systems: Concepts and Techniques* (2nd edn.). Pitman: New York, **1992**.

Billinton, R., & Allan, R. N. *Reliability Evaluation of Power Systems* (2nd edn.). Plenum Press: New York, **1996**.

Boroujeni, H. F., Eghtedari, M., Abdollahi, M., & Behzadipour, E. Calculation of generation system reliability index: Loss of load probability. *Life Science Journal*, **2012**, *9*(4), 4903–4908.

Chaiamarit, K., & Nuchprayoon, S. Modeling of renewable energy resources for generation reliability evaluation. *Renewable and Sustainable Energy Reviews*, **2013**, *26*, 34–41.

Cui, H., Li, F., Fang, X., Chen, H., & Wang, H. Bilevel arbitrage potential evaluation for grid-scale energy storage considering wind power and LMP smoothing effect. *IEEE Transactions on Sustainable Energy*, **2017**.

Davidov, S., & Pantos, M. Optimization model for charging infrastructure planning with electric power system reliability check. *Energy*, **2019**, *166*, 886–894.

Denholm, P., & Mai, T. Timescales of energy storage needed for reducing renewable energy curtailment. *Renewable Energy*, **2019**, *130*, 388–399.

Elsaiah, S., Benidris, M., & Mitra, J. Reliability-constrained optimal distribution system reconfiguration. *Chaos Modeling and Control Systems Design, Studies in Computational Intelligence, 581*, Springer International Publishing, Switzerland, **2015**. doi: 10.1007/978–3–319–13132–0_11.

Florencias-Oliveros, O., González-De-La-Rosa, J., Agüera-Pérez, A., & Palomares-Salas, J. Discussion on reliability and power quality in the smart grid: A prosumer approach of a time scalable index. *International Conference on Renewable Energies and Power Quality (ICREPQ'18)*, Salamanca (Spain), **2018**.

Honrubia-Escribano, A., Giménez, D. U. L., Borroy, V. S., Martín, A. S., & García-Gracia, M. Novel power system reliability indices calculation method. *23rd International Conference on Electricity Distribution, Lyon, Paper 1012*, **2015**.

IEEE Standard 493–2007. *IEEE Recommended Practice for the Design of Reliable Industrial and Commercial Power Systems: Gold Book*, **2007**.

Indiana Utility Regulatory Commission, Reliability Report Data **2002–2009,** Investor-Owned Utilities. Retrieve: www.in.gov/iurc/files/Reliability_Report_Data_2002–2009.pdf (Accessed on 12 October 2019).

Kim, M. C. Reliability block diagram with general gates and its application to system reliability analysis. *Annals of Nuclear Energy*, **2011**, *38*, 2456–2461.

Kumar, T. B., Sekhar, O. C., & Ramamoorthy, M. Composite power system reliability evaluation using modified minimal cut set approach. *Alexandria Engineering Journal*, **2017**, Online.

Li, Y., Miao, S., Zhang, S., Yin, B., Luo, X., Dooner, M., & Wang, J. A reserve capacity model of AA-CAES for power system optimal joint energy and reserve scheduling. *International Journal of Electrical Power and Energy Systems*, **2019**, *104*, 279–290.

Lv, J., Pawlak, M., Annakkage, U. D., & Bagen, B. Statistical testing for load models using measured data. *Electric Power Systems Research*, **2018**, *163*(A), 66–72.

Ma, J., Fouladirad, M., & Grall, A. Flexible wind speed generation model: Markov chain with an embedded diffusion process. *Energy*, **2018**, *164*, 316–328.

Mosaad, S. A. A., Issa, U. H., & Hassan, M. S. Risks affecting the delivery of HVAC systems: Identifying and analysis. *Journal of Building Engineering*, **2018**, *16*, 20–30.

Motalleb, M., Thornton, M., Reihani, E., & Ghorbani, R. A nascent market for contingency reserve services using demand response. *Applied Energy*, **2016**, *179*, 985–995.

Nemeş, C. F., & Munteanu, F. A. Probabilistic model for power generation adequacy evaluation. *Revue Roumaine des Sciences Techniques-Serie Électrotechnique et Énergétique*, **2011**, *56*(1), 36–46.

North American Electric Reliability Corporation (NERC). *Probabilistic Assessment: Technical Guideline Document*, **2016**.

Rebaiaia, K. L., & Ait-Kadi, D. A New technique for generating minimal cut sets in nontrivial network. *AASRI Procedia*, **2013**, *5*, 67–76.

Reimers, A., Cole, W., & Frew, B. The impact of planning reserve margins in long-term planning models of the electricity sector. *Energy Policy*, **2019**, *125*, 1–8.

Resource Adequacy Analysis Subcommittee. 11-year planning horizon: June 1st 2018–May 31st 2029 analysis performed by PJM staff, **2018**. https://www.pjm.com/-/media/committees-groups/subcommittees/raas/20181004/20181004-pjm-reserve-requirement-study-draft-2018.ashx.

Rusin, A., & Wojaczek, A. Trends of changes in the power generation system structure and their impact on the system reliability. *Energy*, **2015**, *92*, 128–134.

Schermeyer, H., Vergara, C., & Fichtner, W. Renewable energy curtailment: A case study on today's and tomorrow's congestion management. *Energy Policy*, **2018**, *112*, 427–436.

Shirvani, M., Memaripour, A., Abdollahi, M., & Salimi, A. Calculation of generation system reliability index: Expected energy not served. *Life Science Journal*, **2012**, *9*(4), 3443–3448.

Staveley-O'Carroll, J. S., & Staveley, O. O. M. International risk sharing in overlapping generations models. *Economics Letters*, **2019**, *174*, 157–160.

Systems Operation Division. *PJM Manual 40: Training and Certification Requirements, Revision 19*, **2018**.

Transmission and Distribution System: Electrical Reliability Annual Report, **2015**. http://www.keysenergy.com/docs/2018-reliability-report.pdf.

Usman, M., Coppo, M., Bignucolo, F., & Turri, R. Losses management strategies in active distribution networks: A review. *Electric Power Systems Research*, **2018**, *163*, 116–132.

Venkatamuni, N. B., & Reddy, B. G. Reliability indices assessment of a small autonomous hybrid power system using aggregate Markov model. *Advanced Research in Electrical and Electronic Engineering*, **2014**, *1*(5), 42–47.

Vivek, V., & Prakash, C. Reliability analysis using renewable energy sources. *International Journal of Mechanical and Industrial Engineering*, **2011**, *1*(1), 47–50.

Wang, Z., Ji, N., Ma, Y., Liu, Z., & Wang, Y. Model selection mechanism of interactive multiple load modeling. *Electrical Power and Energy Systems*, **2018**, *103*, 58–66.

Xie, D., Hui, H., Ding, Y., Lin, Z. Operating reserve capacity evaluation of aggregated heterogeneous TCLs with price signals, *Applied Energy*, Elsevier, **2018**, *216*(C), 338–347.

Yu, J., Guo, L., Ma, M., Kamel, S., Li, W., & Song, X. Risk assessment of integrated electrical, natural gas and district heating systems considering solar thermal CHP plants and electric boilers. *Electrical Power and Energy Systems*, **2018**, *103*, 277–287.

Zhao, Y., Che, Y., Lin, T., Wang, C., Liu, J., Xu, J., & Zhou, J. Minimal cut sets-based reliability evaluation of the more electric aircraft power system. *Hindawi, Mathematical Problems in Engineering*, **2018**, Article ID 9461823, p. 11. https://doi.org/10.1155/2018/9461823.

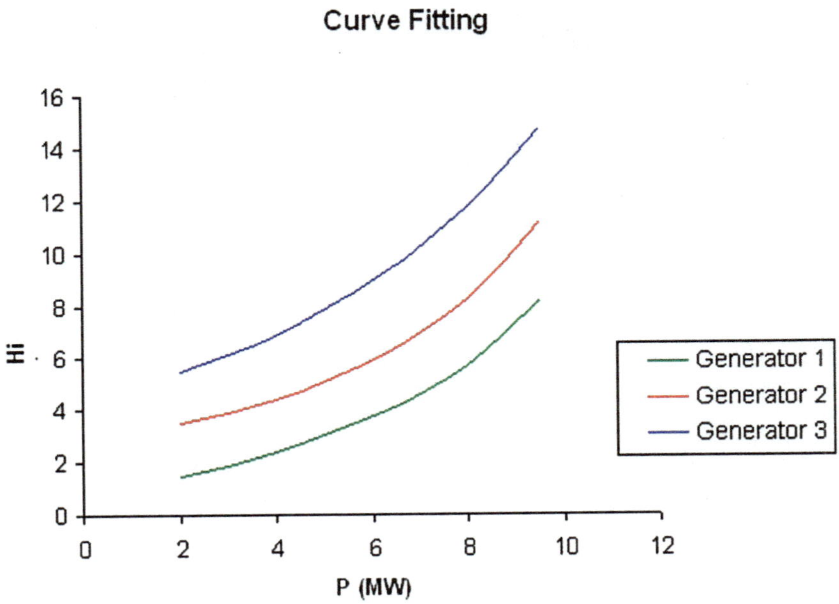

FIGURE 3.2 Curve fitting for the three generators.

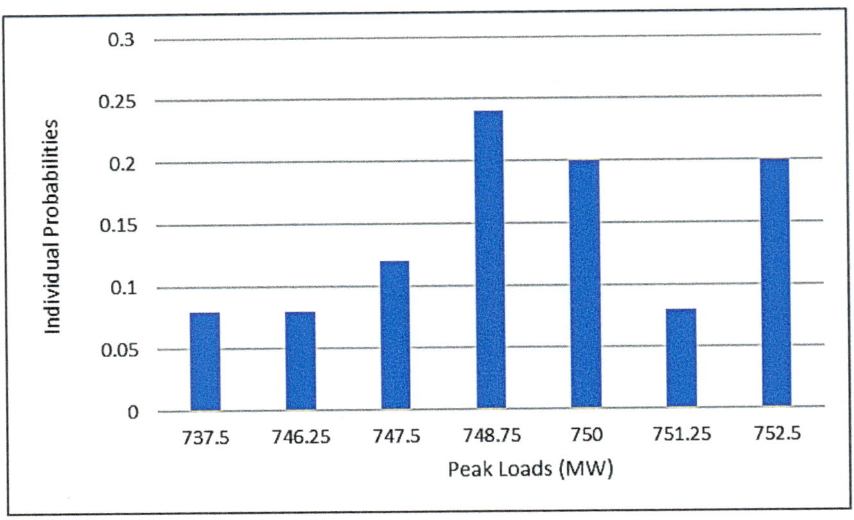

FIGURE 5.10 Individual probabilities vs. peak loads (MW).

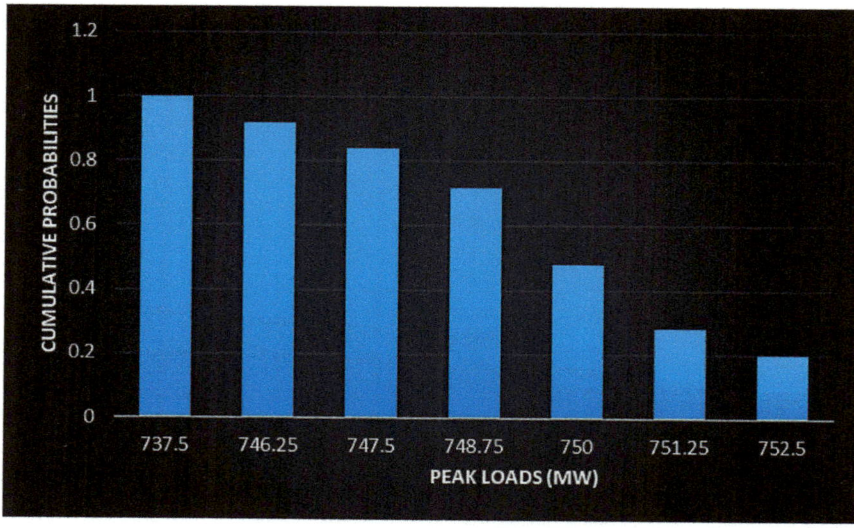

FIGURE 5.11 Cumulative probabilities vs. peak loads (MW).

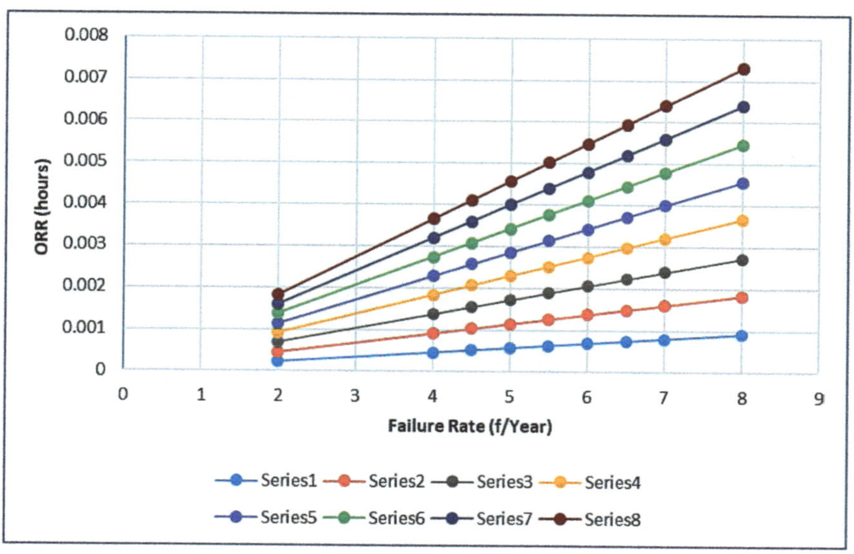

FIGURE 6.9 The final results.

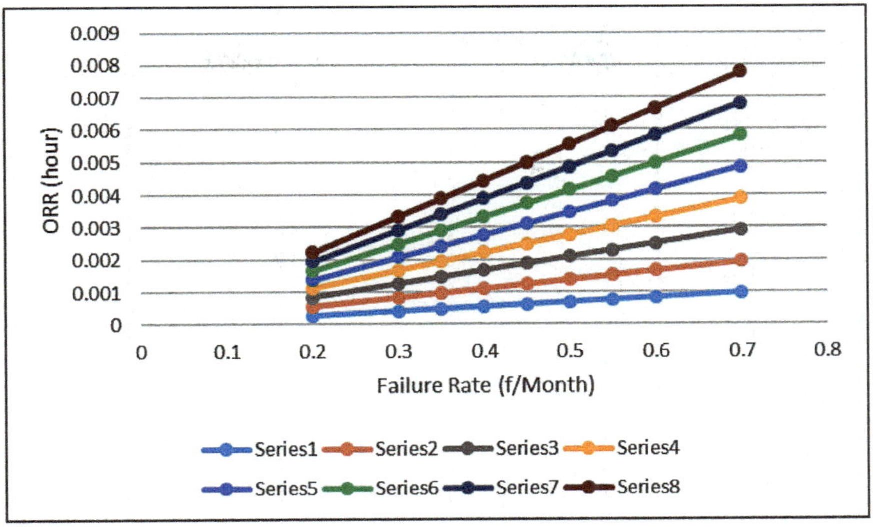

FIGURE 6.10 ORR vs. failure rate relationship.

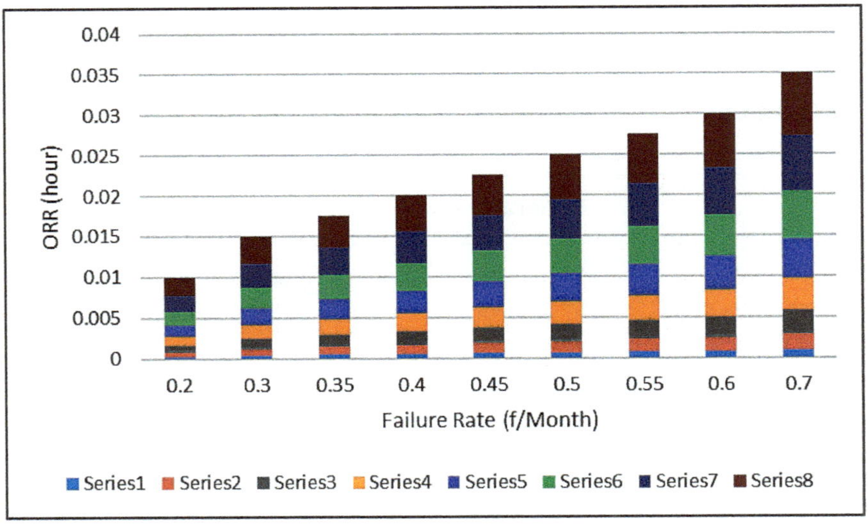

FIGURE 6.11 ORR vs. failure rate histogram.

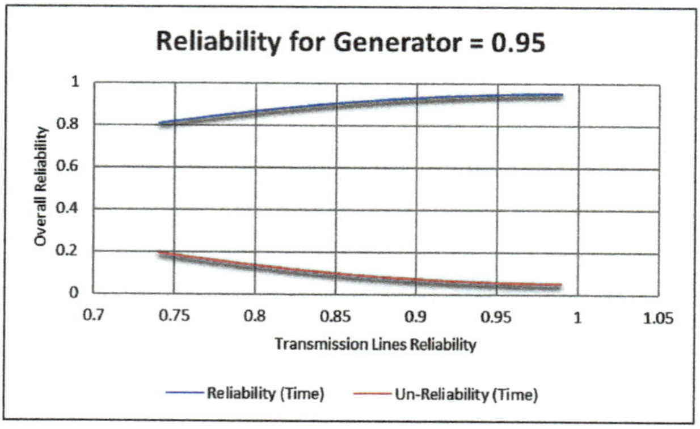

FIGURE 6.22 Overall reliability vs. T.Ls reliability.

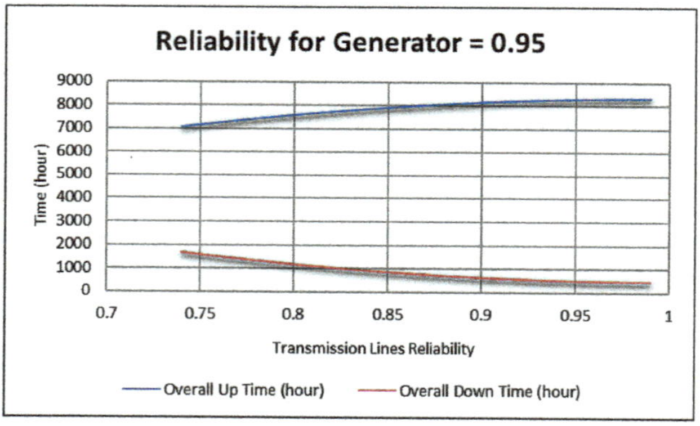

FIGURE 6.23 Time vs. T.Ls reliability.

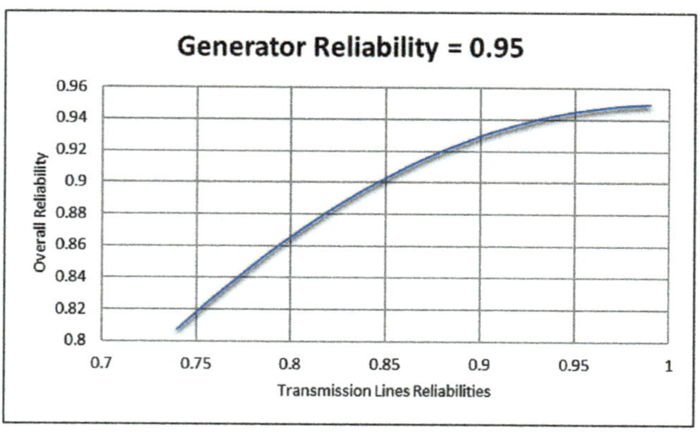

FIGURE 6.24 Overall reliability vs. transmission lines reliabilities.

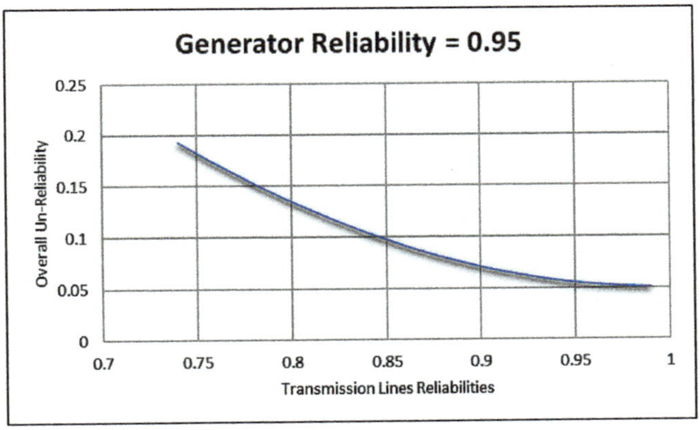

FIGURE 6.25 Overall un-reliability vs. transmission lines reliabilities.

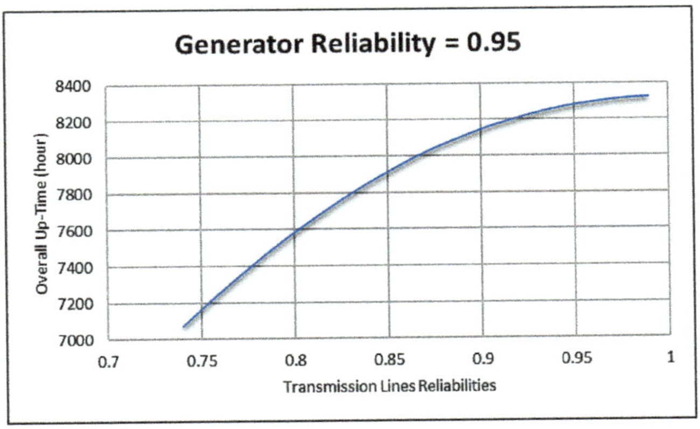

FIGURE 6.26 Overall up-time (hour) vs. transmission lines reliabilities.

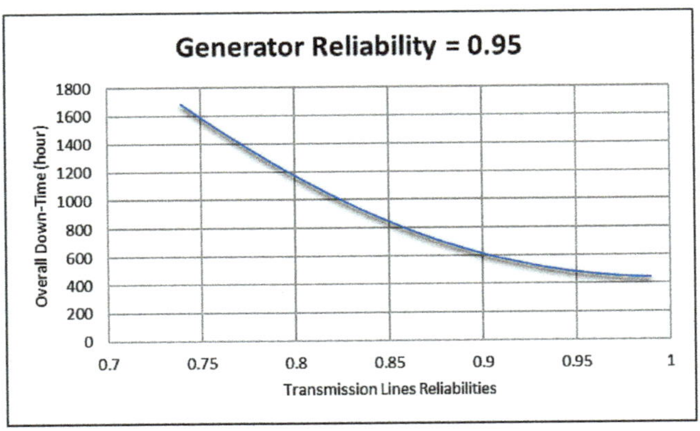

FIGURE 6.27 Overall down-time (hour) vs. transmission lines reliabilities.

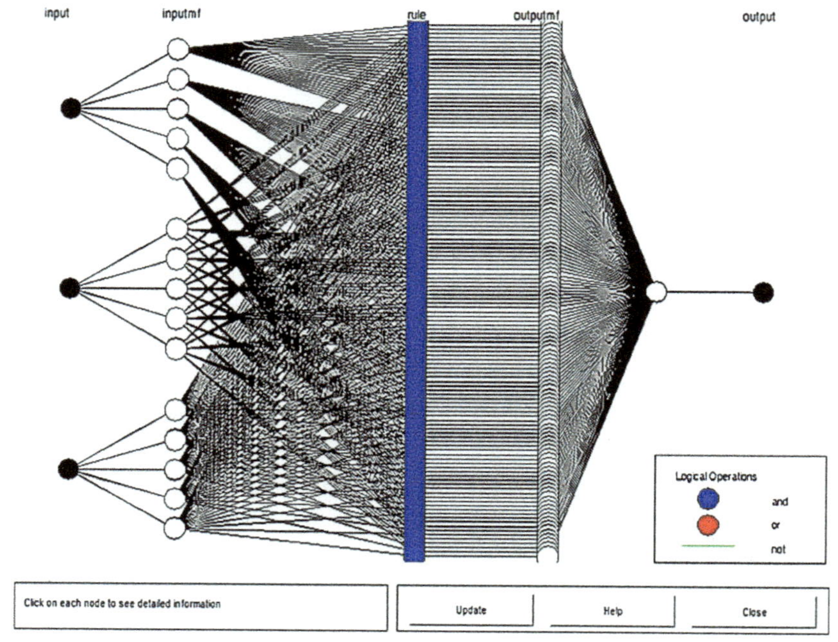

FIGURE 10.14 ANFIS structure with three inputs and one output.
Source: Qamber and Al-Hamad, 2016.

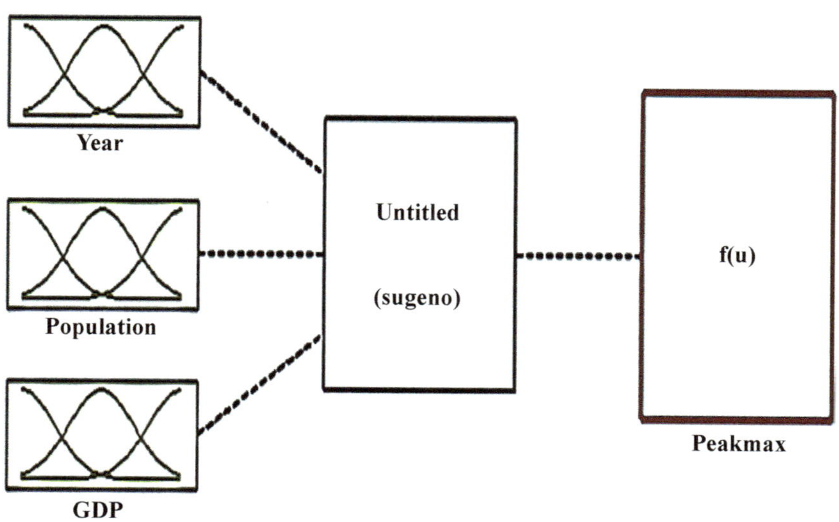

FIGURE 10.15 Artificial neuro-fuzzy logic model using long term load forecasting.
Source: Qamber and Al-Hamad, 2016.

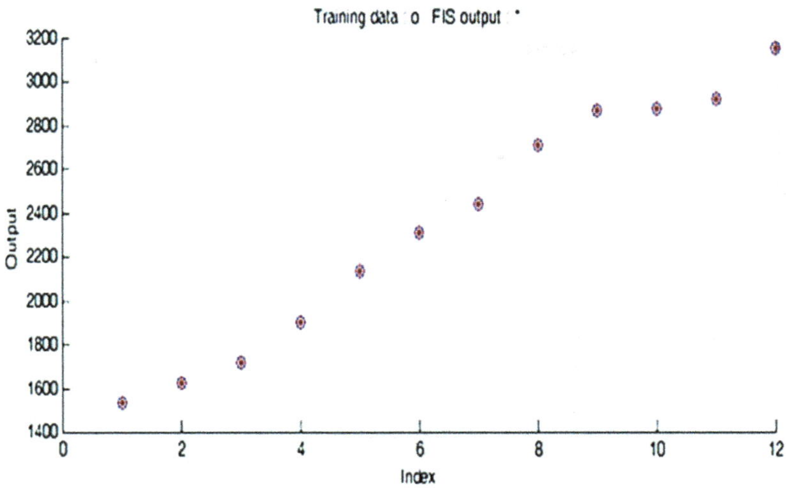

FIGURE 10.16 Actual data used versus the output of the neuro-fuzzy model.
Source: Qamber and Al-Hamad, 2016.

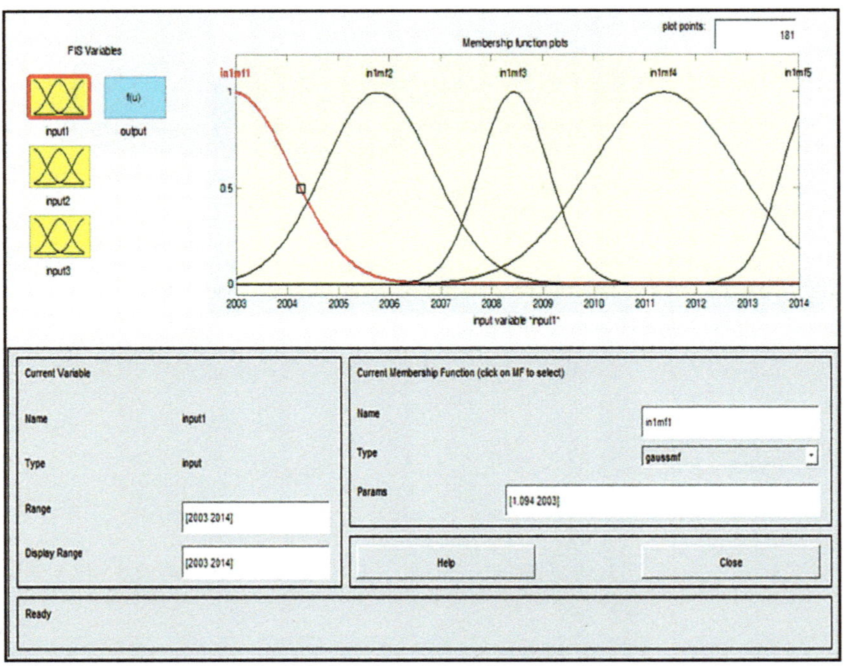

FIGURE 10.17 Membership function used in the neuro-fuzzy logic model for long term load forecasting.
Source: Qamber and Al-Hamad, 2016.

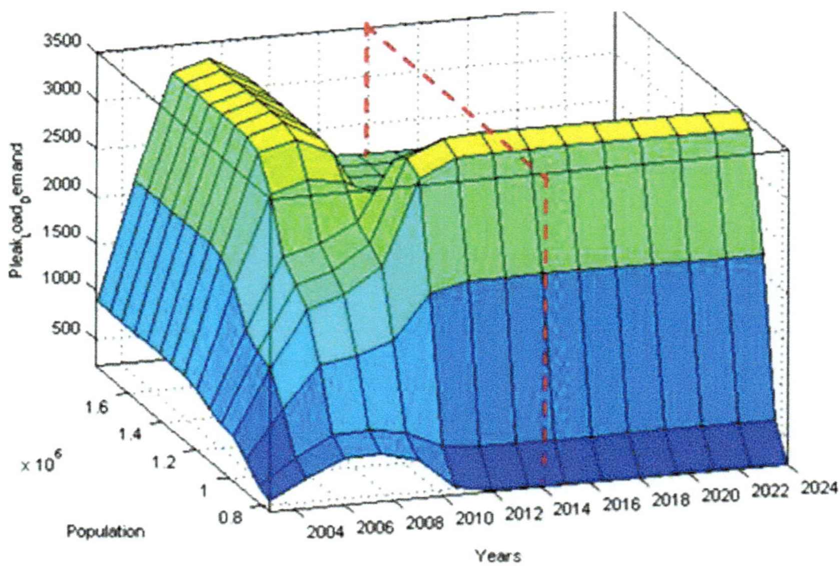

FIGURE 10.18 A surface graph for a country, estimating the load.
Source: Qamber and Al-Hamad, 2016.

CHAPTER 7

Building Power Plant

7.1 INTRODUCTION

Any power plant consists of a number of generators. The present chapter presents each generator with two or more states that the generator is passing through. The generator status represented in a matrix form. To combine the generators, a special type of mathematics needed and illustrated. The matrix dimension is the form based on the generator status. The present chapter starts with the basic definitions and properties of the Kronecker product of matrices. At the same time, it should be noted that the Kronecker products have many applications such as image processing, signal processing, and other applications. In the present chapter, the Kronecker technique presented and applied to the power plant, where it is forming a power station. The Kronecker results compared with direct differentiation results. The conclusion reached with that the Kronecker product is more convenient and simpler than the direct differentiation method (Society for Industrial and Applied Mathematics, 2005).

7.2 KRONECKER MULTIPLICATION

The general rule for Kronecker multiplication shown by considering two subsystems A and B, where the subsystem is illustrated more clear later in the real cases of applications (Shakeri et al., 2016; Mariet et al., 2016; Kepner et al., 2018).

$$\mathrm{A} = \begin{bmatrix} a_{11} & a_{12} \\ a_{21} & a_{22} \end{bmatrix} \text{ and } \mathrm{B} = \begin{bmatrix} b_{11} & b_{12} \\ b_{21} & b_{22} \end{bmatrix}$$

Then multiplication of AB defined as:

$$A \otimes B = \begin{bmatrix} a_{11}B & a_{12}B \\ a_{21}B & a_{22}B \end{bmatrix}$$

where: $\otimes$ defined as Kronecker multiplication sign

$$A \otimes B = \begin{bmatrix} a_{11}\begin{bmatrix} b_{11} & b_{12} \\ b_{21} & b_{22} \end{bmatrix} & a_{12}\begin{bmatrix} b_{11} & b_{12} \\ b_{21} & b_{22} \end{bmatrix} \\ a_{21}\begin{bmatrix} b_{11} & b_{12} \\ b_{21} & b_{22} \end{bmatrix} & a_{22}\begin{bmatrix} b_{11} & b_{12} \\ b_{21} & b_{22} \end{bmatrix} \end{bmatrix}$$

$$= \begin{bmatrix} a_{11}b_{11} & a_{11}b_{12} & a_{12}b_{11} & a_{11}b_{12} \\ a_{11}b_{21} & a_{11}b_{22} & a_{12}b_{21} & a_{12}b_{22} \\ a_{21}b_{11} & a_{21}b_{12} & a_{22}b_{11} & a_{22}b_{12} \\ a_{21}b_{21} & a_{21}b_{22} & a_{22}b_{21} & a_{22}b_{22} \end{bmatrix}$$

In the reliability study, the system is passing through two-state models or more. Therefore, it will generalize as n-number of states that represented as follows:

Let

$$A = \begin{bmatrix} a_{11} & a_{12} & . & . & a_{1n} \\ a_{21} & a_{22} & . & . & a_{2n} \\ . & . & . & . & . \\ . & . & . & . & . \\ a_{n1} & a_{n2} & . & . & a_{nn} \end{bmatrix}$$

and

$$B = \begin{bmatrix} b_{11} & b_{12} & . & . & b_{1n} \\ b_{21} & b_{22} & . & . & b_{2n} \\ . & . & . & . & . \\ . & . & . & . & . \\ b_{n1} & b_{n2} & . & . & b_{nn} \end{bmatrix}$$

Then A ⊗ B is defined as:

$$\begin{bmatrix} a_{11}\begin{bmatrix} b_{11} & b_{12} & . & b_{1n} \\ b_{12} & b_{22} & . & b_{2n} \\ . & . & . & . \\ b_{1n} & b_{2n} & . & b_{nn} \end{bmatrix} & a_{12}\begin{bmatrix} b_{11} & b_{12} & . & b_{1n} \\ b_{12} & b_{22} & . & b_{2n} \\ . & . & . & . \\ b_{1n} & b_{2n} & . & b_{nn} \end{bmatrix} & . \; . & a_{1n}\begin{bmatrix} b_{11} & b_{12} & . & b_{1n} \\ b_{12} & b_{22} & . & b_{2n} \\ . & . & . & . \\ b_{1n} & b_{2n} & . & b_{nn} \end{bmatrix} \\ a_{21}\begin{bmatrix} b_{11} & b_{12} & . & b_{1n} \\ b_{12} & b_{22} & . & b_{2n} \\ . & . & . & . \\ b_{1n} & b_{2n} & . & b_{nn} \end{bmatrix} & a_{22}\begin{bmatrix} b_{11} & b_{12} & . & b_{1n} \\ b_{12} & b_{22} & . & b_{2n} \\ . & . & . & . \\ b_{1n} & b_{2n} & . & b_{nn} \end{bmatrix} & . \; . & a_{2n}\begin{bmatrix} b_{11} & b_{12} & . & b_{1n} \\ b_{12} & b_{22} & . & b_2 n \\ . & . & . & . \\ b_{1n} & b_{2n} & . & b_{nn} \end{bmatrix} \\ . & . & . \; . & . \\ . & . & . \; . & . \\ a_{1n}\begin{bmatrix} b_{11} & b_{12} & . & b_{1n} \\ b_{12} & b_{22} & . & b_{2n} \\ . & . & . & . \\ b_{1n} & b_{2n} & . & b_{nn} \end{bmatrix} & a_{2n}\begin{bmatrix} b_{11} & b_{12} & . & b_{1n} \\ b_{12} & b_{22} & . & b_{2n} \\ . & . & . & . \\ b_{1n} & b_{2n} & . & b_{nn} \end{bmatrix} & . \; . & a_{nn}\begin{bmatrix} b_{11} & b_{12} & . & b_{1n} \\ b_{12} & b_{22} & . & b_{2n} \\ . & . & . & . \\ b_{1n} & b_{2n} & . & b_{nn} \end{bmatrix} \end{bmatrix}.$$

The general shape is as shown above for n number of states for both systems.

7.3 RULES AND PROPERTIES FOR KRONECKER PRODUCTS

Using the above definition of multiplication (Broxson, 2006 and Van Loan, 2000), the ensuring results can be illustrated in the coming sub-sections (Shakeri et al., 2016; Mariet et al., 2016; Kepner et al., 2018).

7.3.1 SCALAR MULTIPLICATION (SHAKERI ET AL., 2016; MARIET ET AL., 2016; KEPNER ET AL., 2018)

In the case of n is a scalar, then

$$A \otimes (nB) = n(A \otimes B) \quad (1)$$

Proof:

The $(i, j)^{th}$ block of $A \otimes (nB)$ is

$$[a_{ij}(nB)] = n\,[a_{ij}B]$$

$$= n[(i, j)^{th} \text{ block of } A \otimes B]$$

7.3.2 MATRIX SUM AND PRODUCTS ASSOCIATION

In the case of matrix summation associated with Kronecker multiplication, the following are the results obtained (Shakeri et al., 2016; Mariet et al., 2016; Kepner et al., 2018):

1. $(A + B) \otimes C = A \otimes C + B \otimes C$ (2)
2. $A \otimes (B + C) = A \otimes B + A \otimes C$ (3)

Proof:

1. The $(i, j)^{th}$ block of LHS $(A + B) \otimes C$ is $(a_{ij} + b_{ij})\,C$

Therefore, the $(i, j)^{th}$ of RHS is

$$A \otimes C + B \otimes C \text{ is}$$

$$A_{ij}\,C + b_{ij}\,C = (a_{ij} + b_{ij})\,C$$

Since the two blocks are equal for every (i, j), the result follows.

2. The $(i, j)^{th}$ block of the LHS $A \otimes (B + C)$ is $A \otimes (b_{ij} + c_{ij})$

Therefore, the $(i, j)^{th}$ block of the other side is

$$A \otimes (b_{ij}) + A \otimes (c_{ij}) = A \otimes (b_{ij} + c_{ij})$$

Since the two blocks are equal for every (i, j), the result follows.

7.3.3 PRODUCT ASSOCIATION

The product of a number of matrices is presented as (Shakeri et al., 2016; Mariet et al., 2016; Kepner et al., 2018):

$$A \otimes (B \otimes C) = (A \otimes B) \otimes C \tag{4}$$

7.3.4 MIXED PRODUCT

The mixed product of a number of matrices is presented as (Shakeri et al., 2016; Mariet et al., 2016; Kepner et al., 2018):

$$(A \otimes B).\,(C \otimes D) = A.C \otimes B.D \tag{5}$$

7.3.5 KRONECKER SUM

Consider the two masteries A(n × n) and B(m × m), then the following (Shakeri et al., 2016; Mariet et al., 2016; Kepner et al., 2018), the Kronecker sum can be defined by the following expression:

$$A \otimes B = A \otimes I_m + I_n \otimes B \tag{6}$$

where:

I_m is a unit matrix (m × m);
I_n is a unit matrix (n × n).

7.4 KRONECKER PRODUCT MODELING

Consider a power station (Shakeri et al., 2016; Broxson, 2006; Mariet et al., 2016; Kepner et al., 2018) with three generating-units as shown in Figure 7.1 and represented by the differential equation describing the subsystem (generating-unit) that can be written as:

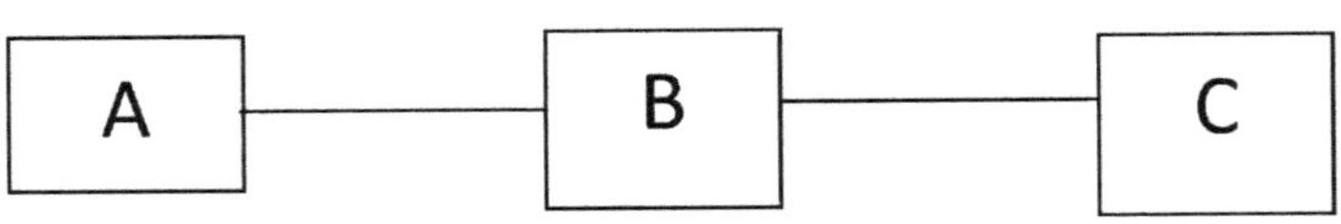

FIGURE 7.1 Three sub-systems (generators).

$$\frac{dP_1(t)}{dt} = T^{(1)}\ \mathrm{P(t)} \quad (1)$$

$$\frac{dP_2(t)}{dt} = T^{(2)}\ \mathrm{P(t)} \quad (2)$$

$$\frac{dP_3(t)}{dt} = T^{(3)}\ \mathrm{P(t)} \quad (3)$$

Let

$$Q(t) = P(t)^{(1)} \otimes P(t)^{(2)} \otimes P(t)^{(3)} \tag{7}$$

where: Q(t) is an equivalent system probability; $T^{(i)}$ is the transition rate matrix for i-th subsystem; $P^{(i)}(t)$ is an i-th column probability vector for the i-th subsystem.

$$\begin{aligned}\frac{dQ(t)}{dx} &= dP(t)^{(1)}/dt \otimes P(t)^{(2)} \otimes P(t)^{(3)} + P(t)^{(1)} \otimes dP(t)^{(2)}/dt \otimes P(t)^{(3)} \\ &\quad + P(t)^{(1)} \otimes P(t)^{(2)} \otimes dP(t)^{(3)}/dt \\ &= T^{(1)}P(t)^{(1)} \otimes P(t)^{(2)} \otimes P(t)^{(3)} + P(t)^{(1)} \otimes T^{(2)}P(t)^{(2)} \otimes P(t)^{(3)} \\ &\quad + P^{(1)} \otimes P(t)^{(2)} \otimes T^{(3)}P(t)^{(3)}\end{aligned}$$

Consider the first term of equation:

$$\begin{aligned}T^{(1)}P(t)^{(1)} \otimes P(t)^{(2)} \otimes P(t)^{(3)} &= [(T^{(1)} \otimes I^{(2)})(P(t)^{(1)} \otimes P(t)^{(2)}] \otimes P(t)^{(3)} \\ &= [T^{(1)} \otimes I^{(2)} \otimes I^{(3)}]\, P(t)^{(1)} \otimes P(t)^{(2)} \otimes P(t)^{(3)} \\ &= [T^{(1)} \otimes I^{(2)} \otimes I^{(3)}]Q(t)\end{aligned}$$

Consider the second term of the equation:

$$\begin{aligned}P(t)^{(1)} \otimes T^{(2)}P(t)^{(2)} \otimes P(t)^{(3)} &= P(t)^{(1)} \otimes (T^{(2)}P(t)^{(2)} \otimes I^{(3)}P(t)^{(3)}) \\ &= P(t)^{(1)} \otimes [(T^{(2)} \otimes I^{(3)})(P(t)^{(2)} \otimes P(t)^{(3)})] \\ &= I^{(1)}P(t)^{(1)} \otimes (T^{(2)} \otimes I^{(3)})(P(t)^{(2)} \otimes P(t)^{(3)}) \\ &= (I^{(1)} \otimes T^{(2)} \otimes I^{(3)})(P(t)^{(1)} \otimes P(t)^{(2)} \otimes P(t)^{(3)}) \\ &= (I^{(1)} \otimes T^{(2)} \otimes I^{(3)})Q(t)\end{aligned}$$

$$\begin{aligned}P(t)^{(1)} \otimes P(t)^{(2)} \otimes T^{(3)}P(t)^{(3)} &= P(t)^{(1)} \otimes [(I^{(2)} \otimes T^{(3)})(P(t)^{(2)} \otimes P(t)^{(3)})] \\ &= (I^{(1)} \otimes I^{(2)}T^{(3)})P(t)^{(1)} \otimes P(t)^{(2)} \otimes P(t)^{(3)} \\ &= (I^{(1)} \otimes I^{(2)} \otimes T^{(3)})Q(t)\end{aligned}$$

Therefore, the equivalent system equation becomes as:

$$\begin{aligned}\frac{dQ(t)}{dt} &= [\{T^{(1)} \otimes I^{(2)} \otimes I^{(3)}\}Q(t) + \{I^{(1)} \otimes T^{(2)} \otimes I^{(3)}Q(t)\} \\ &\quad + \{I^{(1)} \otimes I^{(2)} \otimes T^{(3)}\}Q(t)] \\ &= [\{T^{(1)} \otimes I^{(2)} \otimes I^{(3)}\} + \{I^{(1)} \otimes T^{(2)} \otimes I^{(3)}\} + \{I^{(1)} \otimes I^{(2)} \otimes T^{(3)}\}]Q(t) \\ &= [T^{(1)} \otimes T^{(2)} \otimes T^{(3)}]Q(t)\end{aligned}$$

The summation of the transition rate matrix, the Kronecker summation becomes:

$$T^{(1)} \otimes T^{(2)} \otimes T^{(3)} = \begin{bmatrix} -A_1 & \mu_3 & \mu_2 & 0 & \mu_1 & 0 & 0 & 0 \\ \lambda_3 & -A_2 & 0 & \mu_2 & 0 & \mu_1 & 0 & 0 \\ \lambda_2 & 0 & -A_3 & \mu_3 & 0 & 0 & \mu_1 & 0 \\ 0 & \lambda_2 & \lambda_3 & -A_4 & 0 & 0 & 0 & \mu_1 \\ \lambda_1 & 0 & 0 & 0 & -A_5 & \mu_3 & \mu_2 & 0 \\ 0 & \lambda_1 & 0 & 0 & \lambda_3 & -A_6 & 0 & \mu_2 \\ 0 & 0 & \lambda_1 & 0 & \lambda_2 & 0 & -A_7 & \mu_3 \\ 0 & 0 & 0 & \lambda_1 & 0 & \lambda_2 & \lambda_3 & -A_8 \end{bmatrix}$$

where:

$$A_1 = \lambda_1 + \lambda_2 + \lambda_3$$
$$A_2 = \lambda_1 + \lambda_2 + \mu_3$$
$$A_3 = \mu_1 + \mu_2 + \lambda_3$$
$$A_4 = \lambda_1 + \mu_2 + \mu_3$$
$$A_5 = \mu_1 + \lambda_2 + \lambda_3$$
$$A_6 = \mu_1 + \lambda_2 + \mu_3$$
$$A_7 = \mu_1 + \mu_2 + \lambda_3$$
$$A_8 = \mu_1 + \mu_2 + \mu_3$$

which is identical to the transition rate matrix obtained in Section 7.5.1 by more means that are laborious.

7.5 SYSTEM BUILDING

Systems formed by combining a number of simpler subsystems (Shakeri et al., 2016; Mariet et al., 2016; Kepner et al., 2018). The availability of each component described by

$$\frac{dP(t)}{dt} = A \ \mathrm{P(t)}$$

However, it is often hard to derive an overall formula for the whole system from the equations describing the subsystems. Such knowledge is necessary in order to obtain an understanding of the total system structure and in particular intersection between subsystems. This is mainly attractive if one needs to introduce common mode failures.

The sub-system matrices are the first step to build the overall matrix representing a large system by an iterative procedure, which systematically, adds each sub-system to the other. The equivalent transition matrix formed from the knowledge of each sub-system failure and repair rates.

Two different methods are discussed in the present chapter. These two methods are direct differentiation and the use of the Kronecker product. Both methods are used to obtain the equivalent transition matrix (Broxson, 2006; Van Loan, 2000), which a mathematical representation of a power station in this research.

7.5.1 SYSTEM BUILDING USING KRONECKER TECHNIQUE

As stated earlier, the Kronecker multiplication is a powerful tool to build the transition rate matrix. Therefore, the Kronecker algebra can easily present any large system by combining all the sub-systems under it. Then, using Kronecker algebra in system building, a Matlab function found to use Kronecker algebra. Furthermore, this function is not within the Matlab program yet (Shakeri et al., 2016; Mariet et al., 2016; Kepner et al., 2018).

As mentioned earlier, the Kronecker sum is:

$$A \oplus B = A \otimes I_m + I_n \otimes B \tag{8}$$

where:

I_m is a unit matrix (m × m);
I_n is a unit matric (n × n).

In the system building function, the size of each sub-system is 2 × 2, this came from the equation

$$\frac{dP_1(t)}{dt} = -\lambda P_1(t) + \mu P_2(t)$$

$$\frac{dP_2(t)}{dt} = +\lambda P_1(t) - \mu P_2(t)$$

So each sub-system matrix will be as of:

$$U_n = \begin{bmatrix} -\lambda_n & \mu_n \\ \lambda_n & -\mu_n \end{bmatrix}$$

The size of the unit matrix will be 2 × 2. The overall size of the transition rate matrix can be calculated from the number of sub-systems *(n)* as follows:

Transition rate matrix = 2^n.

Let a system contain two sub-systems A and B. Then:

$$\text{Transition rate matrix} = A \otimes B = A \otimes I_B + I_A \otimes B$$

$$A=\begin{bmatrix} a_{11} & a_{12} \\ a_{21} & a_{22} \end{bmatrix} \text{ and } \quad B=\begin{bmatrix} b_{11} & b_{12} \\ b_{21} & b_{22} \end{bmatrix}$$

Then,

$$A \otimes I_B = \begin{bmatrix} a_{11} & a_{12} \\ a_{21} & a_{22} \end{bmatrix} \otimes \begin{bmatrix} 1 & 0 \\ 0 & 1 \end{bmatrix}$$

and

$$I_A \otimes B = \begin{bmatrix} 1 & 0 \\ 0 & 1 \end{bmatrix} \otimes \begin{bmatrix} b_{11} & b_{12} \\ b_{21} & b_{22} \end{bmatrix}$$

$$A \otimes I_B = \begin{bmatrix} a_{11} & a_{12} \\ a_{21} & a_{22} \end{bmatrix} \otimes \begin{bmatrix} 1 & 0 \\ 0 & 1 \end{bmatrix} = \begin{bmatrix} a_{11}\begin{bmatrix} 1 & 0 \\ 0 & 1 \end{bmatrix} & a_{12}\begin{bmatrix} 1 & 0 \\ 0 & 1 \end{bmatrix} \\ a_{21}\begin{bmatrix} 1 & 0 \\ 0 & 1 \end{bmatrix} & a_{22}\begin{bmatrix} 1 & 0 \\ 0 & 1 \end{bmatrix} \end{bmatrix}$$

$$I_A \otimes B = \begin{bmatrix} 1 & 0 \\ 0 & 1 \end{bmatrix} \otimes \begin{bmatrix} b_{11} & b_{12} \\ b_{21} & b_{22} \end{bmatrix} = \begin{bmatrix} 1\begin{bmatrix} b_{11} & b_{12} \\ b_{21} & b22 \end{bmatrix} & 0\begin{bmatrix} b_{11} & b_{12} \\ b_{21} & b_{22} \end{bmatrix} \\ 0\begin{bmatrix} b_{11} & b_{12} \\ b_{21} & b_{22} \end{bmatrix} & 1\begin{bmatrix} b_{11} & b_{12} \\ b_{21} & b_{22} \end{bmatrix} \end{bmatrix}$$

$$\begin{bmatrix} a_{11}\begin{bmatrix} 1 & 0 \\ 0 & 1 \end{bmatrix} & a_{12}\begin{bmatrix} 1 & 0 \\ 0 & 1 \end{bmatrix} \\ a_{21}\begin{bmatrix} 1 & 0 \\ 0 & 1 \end{bmatrix} & a_{22}\begin{bmatrix} 1 & 0 \\ 0 & 1 \end{bmatrix} \end{bmatrix} = \begin{bmatrix} a_{11} & 0 & a_{12} & 0 \\ 0 & a_{11} & 0 & a_{12} \\ a_{21} & 0 & a_{22} & 0 \\ 0 & a_{21} & 0 & a_{22} \end{bmatrix}$$

$$\begin{bmatrix} 1\begin{bmatrix} b_{11} & b_{12} \\ b_{21} & b_{22} \end{bmatrix} & 0\begin{bmatrix} b_{11} & b_{12} \\ b_{21} & b_{22} \end{bmatrix} \\ 0\begin{bmatrix} b_{11} & b_{12} \\ b_{21} & b_{22} \end{bmatrix} & 1\begin{bmatrix} b_{11} & b_{12} \\ b_{21} & b_{22} \end{bmatrix} \end{bmatrix} = \begin{bmatrix} b_{11} & b_{12} & 0 & 0 \\ b_{21} & b_{22} & 0 & 0 \\ 0 & 0 & b_{11} & b_{12} \\ 0 & 0 & b_{21} & b_{22} \end{bmatrix}$$

Then

$$A \otimes B = A \otimes I_B + I_A \otimes B = \begin{bmatrix} a_{11} & 0 & a_{12} & 0 \\ 0 & a_{11} & 0 & a_{12} \\ a_{21} & 0 & a_{22} & 0 \\ 0 & a_{21} & 0 & a_{22} \end{bmatrix} + \begin{bmatrix} b_{11} & b_{12} & 0 & 0 \\ b_{21} & b_{22} & 0 & 0 \\ 0 & 0 & b_{11} & b_{12} \\ 0 & 0 & b_{21} & b_{22} \end{bmatrix}$$

Then the transition rate matrix for system containing two subsystems A and B are:

$$A \otimes B = A \otimes I_B + I_A \otimes B = \begin{bmatrix} a_{11}+b_{11} & b_{12} & a_{12} & 0 \\ b_{21} & a_{11}+b_{22} & 0 & a_{12} \\ a_{21} & 0 & a_{22}+b_{11} & b_{21} \\ 0 & a_{21} & b_{21} & a_{22}+b_{22} \end{bmatrix}$$

It should be noted that the Kronecker multiplication of a two-unit matrix with the same size is equal to doubling the size of the matrix, i.e., let I_1 size 2×2; I_2 size 2×2 and I_3 size 2×2, then

$$I_1 \otimes I_2 = I \text{ size } 4 \times 4 \qquad (9)$$
$$I_1 \otimes I_2 \otimes I_3 = I \text{ size } 8 \times 8$$

Proof:

$$I_1 \otimes I_2 = \begin{bmatrix} 1 & 0 \\ 0 & 1 \end{bmatrix} \otimes \begin{bmatrix} 1 & 0 \\ 0 & 1 \end{bmatrix} = \begin{bmatrix} 1\begin{bmatrix} 1 & 0 \\ 0 & 1 \end{bmatrix} & 0\begin{bmatrix} 1 & 0 \\ 0 & 1 \end{bmatrix} \\ 0\begin{bmatrix} 1 & 0 \\ 0 & 1 \end{bmatrix} & 1\begin{bmatrix} 1 & 0 \\ 0 & 1 \end{bmatrix} \end{bmatrix} = \begin{bmatrix} 1 & 0 & 0 & 0 \\ 0 & 1 & 0 & 0 \\ 0 & 0 & 1 & 0 \\ 0 & 0 & 0 & 1 \end{bmatrix}$$

$I_1 \otimes I_2 \otimes I_3 =$

$$\begin{bmatrix} 1 & 0 & 0 & 0 \\ 0 & 1 & 0 & 0 \\ 0 & 0 & 1 & 0 \\ 0 & 0 & 0 & 1 \end{bmatrix} \otimes \begin{bmatrix} 1 & 0 \\ 0 & 1 \end{bmatrix} = \begin{bmatrix} 1\begin{bmatrix} 1 & 0 \\ 0 & 1 \end{bmatrix} & 0\begin{bmatrix} 1 & 0 \\ 0 & 1 \end{bmatrix} & 0\begin{bmatrix} 1 & 0 \\ 0 & 1 \end{bmatrix} & 0\begin{bmatrix} 1 & 0 \\ 0 & 1 \end{bmatrix} \\ 0\begin{bmatrix} 1 & 0 \\ 0 & 1 \end{bmatrix} & 1\begin{bmatrix} 1 & 0 \\ 0 & 1 \end{bmatrix} & 0\begin{bmatrix} 1 & 0 \\ 0 & 1 \end{bmatrix} & 0\begin{bmatrix} 1 & 0 \\ 0 & 1 \end{bmatrix} \\ 0\begin{bmatrix} 1 & 0 \\ 0 & 1 \end{bmatrix} & 0\begin{bmatrix} 1 & 0 \\ 0 & 1 \end{bmatrix} & 1\begin{bmatrix} 1 & 0 \\ 0 & 1 \end{bmatrix} & 0\begin{bmatrix} 1 & 0 \\ 0 & 1 \end{bmatrix} \\ 0\begin{bmatrix} 1 & 0 \\ 0 & 1 \end{bmatrix} & 0\begin{bmatrix} 1 & 0 \\ 0 & 1 \end{bmatrix} & 0\begin{bmatrix} 1 & 0 \\ 0 & 1 \end{bmatrix} & 1\begin{bmatrix} 1 & 0 \\ 0 & 1 \end{bmatrix} \end{bmatrix}$$

Then,

$$I_1 \otimes I_2 \otimes I_3 = \begin{bmatrix} 1 & 0 & 0 & 0 & 0 & 0 & 0 & 0 \\ 0 & 1 & 0 & 0 & 0 & 0 & 0 & 0 \\ 0 & 0 & 1 & 0 & 0 & 0 & 0 & 0 \\ 0 & 0 & 0 & 1 & 0 & 0 & 0 & 0 \\ 0 & 0 & 0 & 0 & 1 & 0 & 0 & 0 \\ 0 & 0 & 0 & 0 & 0 & 1 & 0 & 0 \\ 0 & 0 & 0 & 0 & 0 & 0 & 1 & 0 \\ 0 & 0 & 0 & 0 & 0 & 0 & 0 & 1 \end{bmatrix}$$

Three sub-systems A, B, and C then the transition rate matrix equal to:

$$A \otimes B \otimes C = A \otimes I_B \otimes I_C + I_B \otimes B \otimes I_C + I_A \otimes I_B \otimes C$$

It is well known that the Kronecker Multiplication step:

$A \otimes I_B \otimes I_C = A \otimes I$ (where I size 4×4)

$I_B \otimes B \otimes I_C =$

$$\begin{bmatrix} a_{11} & a_{12} \\ a_{21} & a_{22} \end{bmatrix} \otimes \begin{bmatrix} 1 & 0 & 0 & 0 \\ 0 & 1 & 0 & 0 \\ 0 & 0 & 1 & 0 \\ 0 & 0 & 0 & 1 \end{bmatrix} = \begin{bmatrix} a_{11}\begin{bmatrix} 1 & 0 & 0 & 0 \\ 0 & 1 & 0 & 0 \\ 0 & 0 & 1 & 0 \\ 0 & 0 & 0 & 1 \end{bmatrix} & a_{12}\begin{bmatrix} 1 & 0 & 0 & 0 \\ 0 & 1 & 0 & 0 \\ 0 & 0 & 1 & 0 \\ 0 & 0 & 0 & 1 \end{bmatrix} \\ a_{21}\begin{bmatrix} 1 & 0 & 0 & 0 \\ 0 & 1 & 0 & 0 \\ 0 & 0 & 1 & 0 \\ 0 & 0 & 0 & 1 \end{bmatrix} & a_{22}\begin{bmatrix} 1 & 0 & 0 & 0 \\ 0 & 1 & 0 & 0 \\ 0 & 0 & 1 & 0 \\ 0 & 0 & 0 & 1 \end{bmatrix} \end{bmatrix}$$

$$= \begin{bmatrix} a_{11} & 0 & 0 & 0 & a_{12} & 0 & 0 & 0 \\ 0 & a_{11} & 0 & 0 & 0 & a_{12} & 0 & 0 \\ 0 & 0 & a_{11} & 0 & 0 & 0 & a_{12} & 0 \\ 0 & 0 & 0 & a_{11} & 0 & 0 & 0 & a_{12} \\ a_{21} & 0 & 0 & 0 & a_{22} & 0 & 0 & 0 \\ 0 & a_{21} & 0 & 0 & 0 & a_{22} & 0 & 0 \\ 0 & 0 & a_{21} & 0 & 0 & 0 & a_{22} & 0 \\ 0 & 0 & 0 & a_{21} & 0 & 0 & 0 & a_{22} \end{bmatrix}$$

$$I_B \otimes B \otimes I_C = \begin{bmatrix} 1 & 0 \\ 0 & 1 \end{bmatrix} \otimes \begin{bmatrix} b_{11} & b_{12} \\ b_{21} & b_{22} \end{bmatrix} \otimes \begin{bmatrix} 1 & 0 \\ 0 & 1 \end{bmatrix}$$

$$= \begin{bmatrix} 1\begin{bmatrix} b_{11} & b_{12} \\ b_{21} & b_{22} \end{bmatrix} & 0\begin{bmatrix} b_{11} & b_{12} \\ b_{21} & b_{22} \end{bmatrix} \\ 0\begin{bmatrix} b_{11} & b_{12} \\ b_{21} & b_{22} \end{bmatrix} & 1\begin{bmatrix} b_{11} & b_{12} \\ b_{21} & b_{22} \end{bmatrix} \end{bmatrix} \otimes \begin{bmatrix} 1 & 0 \\ 0 & 1 \end{bmatrix} = \begin{bmatrix} b_{11} & b_{12} & 0 & 0 \\ b_{21} & b_{22} & 0 & 0 \\ 0 & 0 & b_{11} & b_{12} \\ 0 & 0 & b_{21} & b_{22} \end{bmatrix} \otimes \begin{bmatrix} 1 & 0 \\ 0 & 1 \end{bmatrix}$$

$$= \begin{bmatrix} b_{11}\begin{bmatrix} 1 & 0 \\ 0 & 1 \end{bmatrix} & b_{12}\begin{bmatrix} 1 & 0 \\ 0 & 1 \end{bmatrix} & 0\begin{bmatrix} 1 & 0 \\ 0 & 1 \end{bmatrix} & 0\begin{bmatrix} 1 & 0 \\ 0 & 1 \end{bmatrix} \\ b_{21}\begin{bmatrix} 1 & 0 \\ 0 & 1 \end{bmatrix} & b_{22}\begin{bmatrix} 1 & 0 \\ 0 & 1 \end{bmatrix} & 0\begin{bmatrix} 1 & 0 \\ 0 & 1 \end{bmatrix} & 0\begin{bmatrix} 1 & 0 \\ 0 & 1 \end{bmatrix} \\ 0\begin{bmatrix} 1 & 0 \\ 0 & 1 \end{bmatrix} & 0\begin{bmatrix} 1 & 0 \\ 0 & 1 \end{bmatrix} & b_{11}\begin{bmatrix} 1 & 0 \\ 0 & 1 \end{bmatrix} & b_{12}\begin{bmatrix} 1 & 0 \\ 0 & 1 \end{bmatrix} \\ 0\begin{bmatrix} 1 & 0 \\ 0 & 1 \end{bmatrix} & 0\begin{bmatrix} 1 & 0 \\ 0 & 1 \end{bmatrix} & b_{21}\begin{bmatrix} 1 & 0 \\ 0 & 1 \end{bmatrix} & b_{22}\begin{bmatrix} 1 & 0 \\ 0 & 1 \end{bmatrix} \end{bmatrix}$$

$$= \begin{bmatrix} b_{11} & 0 & b_{12} & 0 & 0 & 0 & 0 & 0 \\ 0 & b_{11} & 0 & b_{12} & 0 & 0 & 0 & 0 \\ b_{21} & 0 & b_{22} & 0 & 0 & 0 & 0 & 0 \\ 0 & b_{21} & 0 & b_{22} & 0 & 0 & 0 & 0 \\ 0 & 0 & 0 & 0 & b_{11} & 0 & b_{12} & 0 \\ 0 & 0 & 0 & 0 & 0 & b_{11} & 0 & b_{12} \\ 0 & 0 & 0 & 0 & b_{21} & 0 & b_{22} & 0 \\ 0 & 0 & 0 & 0 & 0 & b_{21} & 0 & b_{22} \end{bmatrix}$$

$$I_A \otimes I_B \otimes C = \begin{bmatrix} 1 & 0 & 0 & 0 \\ 0 & 1 & 0 & 0 \\ 0 & 0 & 1 & 0 \\ 0 & 0 & 0 & 1 \end{bmatrix} \otimes \begin{bmatrix} c_{11} & c_{12} \\ c_{21} & c_{22} \end{bmatrix}$$

$$= \begin{bmatrix} 1\begin{bmatrix} c_{11} & c_{12} \\ c_{21} & c_{22} \end{bmatrix} & 0\begin{bmatrix} c_{11} & c_{12} \\ c_{21} & c_{22} \end{bmatrix} & 0\begin{bmatrix} c_{11} & c_{12} \\ c_{21} & c_{22} \end{bmatrix} & 0\begin{bmatrix} c_{11} & c_{12} \\ c_{21} & c_{22} \end{bmatrix} \\ 0\begin{bmatrix} c_{11} & c_{12} \\ c_{21} & c_{22} \end{bmatrix} & 1\begin{bmatrix} c_{11} & c_{12} \\ c_{21} & c_{22} \end{bmatrix} & 0\begin{bmatrix} c_{11} & c_{12} \\ c_{21} & c_{22} \end{bmatrix} & 0\begin{bmatrix} c_{11} & c_{12} \\ c_{21} & c_{22} \end{bmatrix} \\ 0\begin{bmatrix} c_{11} & c_{12} \\ c_{21} & c_{22} \end{bmatrix} & 0\begin{bmatrix} c_{11} & c_{12} \\ c_{21} & c_{22} \end{bmatrix} & 1\begin{bmatrix} c_{11} & c_{12} \\ c_{21} & c_{22} \end{bmatrix} & 0\begin{bmatrix} c_{11} & c_{12} \\ c_{21} & c_{22} \end{bmatrix} \\ 0\begin{bmatrix} c_{11} & c_{12} \\ c_{21} & c_{22} \end{bmatrix} & 0\begin{bmatrix} c_{11} & c_{12} \\ c_{21} & c_{22} \end{bmatrix} & 0\begin{bmatrix} c_{11} & c_{12} \\ c_{21} & c_{22} \end{bmatrix} & 1\begin{bmatrix} c_{11} & c_{12} \\ c_{21} & c_{22} \end{bmatrix} \end{bmatrix}$$

$$= \begin{bmatrix} c_{11} & c_{12} & 0 & 0 & 0 & 0 & 0 & 0 \\ c_{21} & c_{22} & 0 & 0 & 0 & 0 & 0 & 0 \\ 0 & 0 & c_{11} & c_{12} & 0 & 0 & 0 & 0 \\ 0 & 0 & c_{21} & c_{22} & 0 & 0 & 0 & 0 \\ 0 & 0 & 0 & 0 & c_{11} & c_{12} & 0 & 0 \\ 0 & 0 & 0 & 0 & c_{21} & c_{22} & 0 & 0 \\ 0 & 0 & 0 & 0 & 0 & 0 & c_{11} & c_{12} \\ 0 & 0 & 0 & 0 & 0 & 0 & c_{21} & c_{22} \end{bmatrix}$$

Therefore, the Kronecker summation

$$A \oplus B \oplus C = A \otimes I_B \otimes I_C + I_B \otimes B \otimes I_C + I_A \otimes I_B \otimes C$$

$$= \begin{bmatrix} a_{11} & 0 & 0 & 0 & a_{12} & 0 & 0 & 0 \\ 0 & a_{11} & 0 & 0 & 0 & a_{12} & 0 & 0 \\ 0 & 0 & a_{11} & 0 & 0 & 0 & a_{12} & 0 \\ 0 & 0 & 0 & a_{11} & 0 & 0 & 0 & a_{12} \\ a_{21} & 0 & 0 & 0 & a_{22} & 0 & 0 & 0 \\ 0 & a_{21} & 0 & 0 & 0 & a_{22} & 0 & 0 \\ 0 & 0 & a_{21} & 0 & 0 & 0 & a_{22} & 0 \\ 0 & 0 & 0 & a_{21} & 0 & 0 & 0 & a_{22} \end{bmatrix} + \begin{bmatrix} b_{11} & 0 & b_{12} & 0 & 0 & 0 & 0 & 0 \\ 0 & b_{11} & 0 & b_{12} & 0 & 0 & 0 & 0 \\ b_{21} & 0 & b_{22} & 0 & 0 & 0 & 0 & 0 \\ 0 & b_{21} & 0 & b_{22} & 0 & 0 & 0 & 0 \\ 0 & 0 & 0 & 0 & b_{11} & 0 & b_{12} & 0 \\ 0 & 0 & 0 & 0 & 0 & b_{11} & 0 & b_{12} \\ 0 & 0 & 0 & 0 & b_{21} & 0 & b_{22} & 0 \\ 0 & 0 & 0 & 0 & 0 & b_{21} & 0 & b_{22} \end{bmatrix}$$

$$+ \begin{bmatrix} c_{11} & c_{12} & 0 & 0 & 0 & 0 & 0 & 0 \\ c_{21} & c_{22} & 0 & 0 & 0 & 0 & 0 & 0 \\ 0 & 0 & c_{11} & c_{12} & 0 & 0 & 0 & 0 \\ 0 & 0 & c_{21} & c_{22} & 0 & 0 & 0 & 0 \\ 0 & 0 & 0 & 0 & c_{11} & c_{12} & 0 & 0 \\ 0 & 0 & 0 & 0 & c_{21} & c_{22} & 0 & 0 \\ 0 & 0 & 0 & 0 & 0 & 0 & c_{11} & c_{12} \\ 0 & 0 & 0 & 0 & 0 & 0 & c_{21} & c_{22} \end{bmatrix} = \begin{bmatrix} z_1 & c_{12} & b_{12} & 0 & a_{12} & 0 & 0 & 0 \\ c_{21} & z_2 & 0 & b_{12} & 0 & a_{12} & 0 & 0 \\ b_{21} & 0 & z_3 & c_{12} & 0 & 0 & a_{12} & 0 \\ 0 & b_{21} & 0 & z_4 & 0 & 0 & 0 & a_{12} \\ a_{21} & 0 & 0 & 0 & z_5 & c_{12} & 0 & 0 \\ 0 & a_{21} & 0 & 0 & c_{21} & z_6 & 0 & 0 \\ 0 & 0 & a_{21} & 0 & b_{21} & 0 & z_7 & c_{12} \\ 0 & 0 & 0 & a_{21} & 0 & b_{21} & c_{21} & z_8 \end{bmatrix}$$

where:

$$z_1 = a_{11} + b_{11} + c_{11}$$
$$z_2 = a_{11} + b_{11} + c_{22}$$
$$z_3 = a_{11} + b_{22} + c_{11}$$
$$z_4 = a_{11} + b_{22} + c_{22}$$
$$z_5 = a_{22} + b_{11} + c_{11}$$
$$z_6 = a_{22} + b_{11} + c_{22}$$
$$z_7 = a_{22} + b_{22} + c_{11}$$
$$z_8 = a_{22} + b_{22} + c_{22}$$

From the above examples, the Kronecker Sum can be generalized in the following equation:

$$A_1 \oplus A_2 \oplus \ldots\ldots \oplus A_n = A_1 \otimes I_1 \otimes I_2 \ldots \otimes I_n + I_1 \otimes A_2 \otimes I_3 \ldots \otimes I_n + I_1 \otimes I_2 \otimes \ldots I_{n-1} \otimes A_n$$

In case, the evaluation of electric power plant is required. Therefore, the Kronecker is applied. For the different electric power plant, the reliability evaluation is required. As an example, five generators are assumed, and their data represented in Table 7.1. The results of combinations for the five generators are 32 steady-state probabilities are calculated and recorded in Table 7.2, where each generating-unit assumed to pass through two states. It should be noted that (1) means the generating-unit is ON and (0) means the generating-unit is OFF. Furthermore, it is assumed that each generator is working with its full capacity.

The calculation of the number of states that the five generators are passing through can be calculated as:

$$\textit{Number of States that system passing through (Five Connected Generators)} = 2^5$$

where:

2 represent the number of states that each generator is passing through;

5 represent the number of generators.

TABLE 7.1 Five Generating-Units (Assumed as First Electric Power Plant)

Unit No.	Capacity (MW)	Number of Failures (Times per year)	MTTR (Hour)
1	100	3	3.770
2	100	5	9.238
3	100	6	4.045
4	100	5	6.844
5	100	4	0.570

TABLE 7.2 Five Generating-Units Results (First Electric Power Plant)

Probability #	Unit # 1	Unit # 2	Unit # 3	Unit # 4	Unit # 5
1	1	1	1	1	1
2	1	1	1	1	0
3	1	1	1	0	1
4	1	1	1	0	0
5	1	1	0	1	1
6	1	1	0	1	0
7	1	1	0	0	1
8	1	1	0	0	0
9	1	0	1	1	1
10	1	0	1	1	0
11	1	0	1	0	1
12	1	0	1	0	0
13	1	0	0	1	1
14	1	0	0	1	0
15	1	0	0	0	1
16	1	0	0	0	0
17	0	1	1	1	1
18	0	1	1	1	0
19	0	1	1	0	1
20	0	1	1	0	0
21	0	1	0	1	1
22	0	1	0	1	0
23	0	1	0	0	1
24	0	1	0	0	0
25	0	0	1	1	1
26	0	0	1	1	0
27	0	0	1	0	1
28	0	0	1	0	0
29	0	0	0	1	1
30	0	0	0	1	0
31	0	0	0	0	1
32	0	0	0	0	0

Then, consider a second power plant representing six generating-units illustrated in Table 7.3. The results of combinations for the six generating-units are 64 steady-state probabilities are calculated and recorded in Table 7.4, where each generating-unit assumed that it is passing through two states. It should be noted that (1) means the generating-unit is ON and (0) means the generating-unit is OFF. Furthermore, it is assumed that each generator is working with its full capacity.

TABLE 7.3 Six Generating-Units (Assumed as Second Electric Power Plant)

Unit No.	Capacity (MW)	Number of Failures (Times per year)	MTTR (Hour)
1	200	3	2.690
2	200	2	1.915
3	200	2	4.010
4	200	1	3.920
5	200	2	2.040
6	200	2	4.300

TABLE 7.4 Six Generating-Units Results (Second Electric Power Plant)

Probability	Unit # 1	Unit # 2	Unit # 3	Unit # 4	Unit # 5	Unit # 6
1	1	1	1	1	1	1
2	1	1	1	1	1	0
3	1	1	1	1	0	1
4	1	1	1	1	0	0
5	1	1	1	0	1	1
6	1	1	1	0	1	0
7	1	1	1	0	0	1
8	1	1	1	0	0	0
9	1	1	0	1	1	1
10	1	1	0	1	1	0
11	1	1	0	1	0	1
12	1	1	0	1	0	0
13	1	1	0	0	1	1
14	1	1	0	0	1	0
15	1	1	0	0	0	1
16	1	1	0	0	0	0
17	1	0	1	1	1	1
18	1	0	1	1	1	0
19	1	0	1	1	0	1
20	1	0	1	1	0	0
21	1	0	1	0	1	1
22	1	0	1	0	1	0
23	1	0	1	0	0	1

TABLE 7.4 *(Continued)*

Probability	Unit # 1	Unit # 2	Unit # 3	Unit # 4	Unit # 5	Unit # 6
24	1	0	1	0	0	0
25	1	0	0	1	1	1
26	1	0	0	1	1	0
27	1	0	0	1	0	1
28	1	0	0	1	0	0
29	1	0	0	0	1	1
30	1	0	0	0	1	0
31	1	0	0	0	0	1
32	1	0	0	0	0	0
33	0	1	1	1	1	1
34	0	1	1	1	1	0
35	0	1	1	1	0	1
36	0	1	1	1	0	0
37	0	1	1	0	1	1
38	0	1	1	0	1	0
39	0	1	1	0	0	1
40	0	1	1	0	0	0
41	0	1	0	1	1	1
42	0	1	0	1	1	0
43	0	1	0	1	0	1
44	0	1	0	1	0	0
45	0	1	0	0	1	1
46	0	1	0	0	1	0
47	0	1	0	0	0	1
48	0	1	0	0	0	0
49	0	0	1	1	1	1
50	0	0	1	1	1	0
51	0	0	1	1	0	1
52	0	0	1	1	0	0
53	0	0	1	0	1	1
54	0	0	1	0	1	0
55	0	0	1	0	0	1
56	0	0	1	0	0	0
57	0	0	0	1	1	1
58	0	0	0	1	1	0
59	0	0	0	1	0	1
60	0	0	0	1	0	0
61	0	0	0	0	1	1
62	0	0	0	0	1	0
63	0	0	0	0	0	1
64	0	0	0	0	0	0

7.5.2 SYSTEM BUILDING USING DIFFERENTIATION TECHNIQUE

The problem that is studied in this section is that of discussing the availability of the system of sub-systems in Figure 7.2, where r = 1, 2, ……, n represents a sub-system r. It is assumed that each sub-system $S^{(r)}$ is either in a state $S_1^{(t)}$, where the sub-system is working or dichotomously in second state $S_2^{(t)}$, where it failed. The probability of being $S_1^{(t)}$ and $S_2^{(t)}$ are given $P_1(t)^{(r)}$ and $P_2(t)^{(r)}$, respectively.

The basic equation for the sub-system is further assumed to be:

$$\left.\begin{aligned}
dP_1(t)^{(1)}/dt &= -\lambda_1 P_1(t)^{(1)} + \mu_1 P_2(t)^{(1)} \\
dP_2(t)^{(1)}/dt &= \lambda_1 P_1(t)^{(1)} - \mu_1 P_2(t)^{(1)} \\
\\
dP_1(t)^{(2)}/dt &= -\lambda_2 P_1(t)^{(2)} + \mu_2 P_2(t)^{(2)} \\
dP_2(t)^{(2)}/dt &= \lambda_2 P_1(t)^{(2)} - \mu_2 P_2(t)^{(2)} \\
&\cdot \\
&\cdot \\
&\cdot \\
dP_1(t)^{(n)}/dt &= -\lambda_1 P_1(t)^{(n)} + \mu_1 P_2(t)^{(n)} \\
dP_2(t)^{(n)}/dt &= \lambda_1 P_1(t)^{(n)} - \mu_1 P_2(t)^{(n)}
\end{aligned}\right\}$$

The overall availability of the system depicted in Figure 7.2 and can be written as:

$$A = \prod_{n=1}^{N} A_n$$

where $A_n = P_1(t)^{(n)}$. These equations can be obtained by one of the two methods; the first methods is the use of differentiation and the second method is the generalized form using Kronecker product of matrices, these two methods are now described in detail (Shakeri et al., 2016; Mariet et al., 2016; Kepner et al., 2018).

To demonstrate the method and it is convenient to exemplify if first two sub-systems in Figure 7.2, each of them, which can be in one of two states given by the equation:

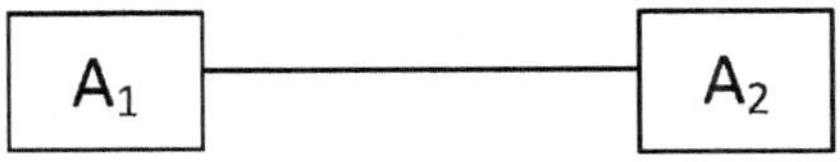

FIGURE 7.2a Two sub-systems generators.

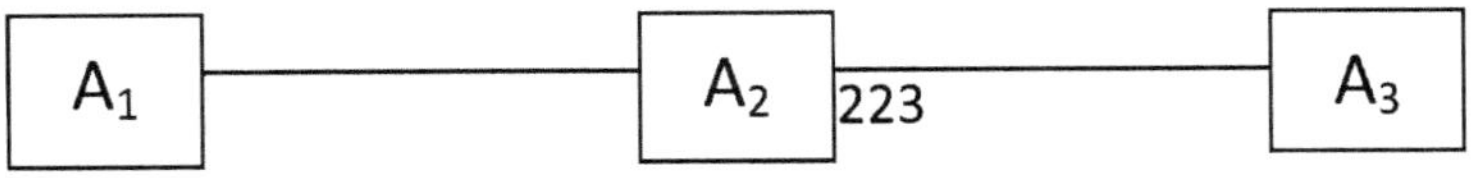

FIGURE 7.2b Three Sub-Systems (Generators)

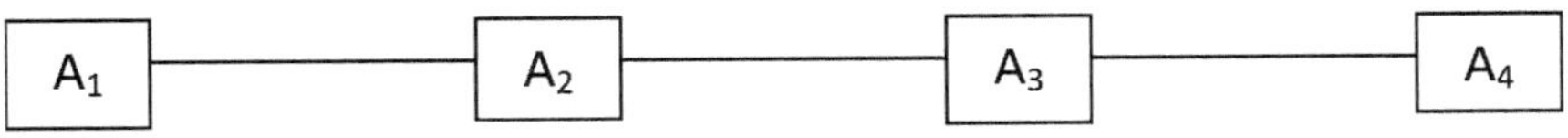

FIGURE 7.2c Four Sub-Systems (Generators)

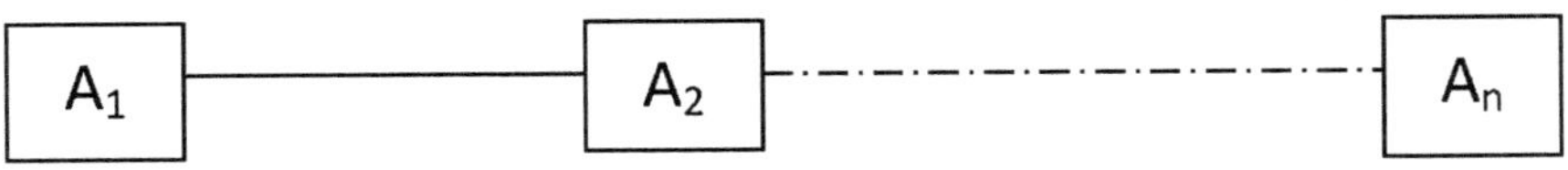

FIGURE 7.2d n Sub-Systems (Generators)

$$\left.\begin{aligned} dP_1(\text{t})^{(1)}/dt &= -\lambda_1 P_1(\text{t})^{(1)} + \mu_1 P_2(\text{t})^{(1)} \\ dP_2(\text{t})^{(1)}/dt &= \lambda_1 P(\text{t})_1^{(1)} - \mu_1 P_2(\text{t})^{(1)} \\ \\ dP_1(\text{t})^{(2)}/dt &= -\lambda_2 P_1(\text{t})^{(2)} + \mu_2 P_2(\text{t})^{(2)} \\ dP_2(\text{t})^{(2)}/dt &= \lambda_2 P_1(\text{t})^{(2)} - \mu_2 P_2(\text{t})^{(2)} \end{aligned}\right\}$$

$$\begin{aligned} dP_1(\text{t})^{(2)}/dt &= -\lambda_2 P_1(\text{t})^{(2)} + \mu_2 P_2(\text{t})^{(2)} \\ dP_2(\text{t})^{(2)}/dt &= \lambda_2 P_1(\text{t})^{(2)} - \mu_2 P_2(\text{t})^{(2)} \end{aligned}$$

Let

$$\left.\begin{aligned} \text{Q}_1(\text{t}) &= \text{P}_1(\text{t})^{(1)}\,\text{P}_1(\text{t})^{(2)} \\ \text{Q}_2(\text{t}) &= \text{P}_1(\text{t})^{(1)}\,\text{P}_2(\text{t})^{(2)} \\ \text{Q}_3(\text{t}) &= \text{P}_2(\text{t})^{(1)}\,\text{P}_1(\text{t})^{(2)} \\ \text{Q}_4(\text{t}) &= \text{P}_2(\text{t})^{(1)}\,\text{P}_2(\text{t})^{(2)} \end{aligned}\right\}$$

Then, differentiating the Q's equations gives,

$$\frac{dQ_1(t)}{dt} = P_1(t)^{(1)}\, dP_1(t)^{(2)}/dt + P_1(t)^{(2)}\, dP_1(t)^{(1)}/dt$$

$$=P_1(t)^{(1)}\,[-\lambda_2 P_1(t)^{(2)} + \lambda_2 P_2(t)^{(2)}] + P_1(t)^{(2)}\,[-\lambda_1 P_1(t)^{(1)} + \lambda_1 P_2(t)^{(1)}]$$

$$= -\lambda_2\, P_1(t)^{(1)}\, P_1(t)^{(2)} + \mu_2\, P_1(t)^{(1)}\, P_2(t)^{(2)} - \lambda_1 P_1(t)^{(1)}\, P_1(t)^{(2)} + \mu_1 P_2(t)^{(1)}\, P_1(t)^{(2)}$$

$$= -\lambda_2\, Q_1(t) + \mu_2\, Q_2(t) - \mu_1 Q_1(t) + \mu_1 Q_3(t)$$

$$= -(\lambda_1 + \lambda_2)\, Q1(t) + \mu_2\, Q_2(t) + \mu_1 Q_3(t)$$

$$\frac{dQ_2(t)}{dt} = P_1^{(1)}(t)\, dP_2(t)^{(2)}/dt + P_2(t)^{(2)}\, dP_1(t)^{(1)}/dt$$

$$= P_1(t)^{(1)}\,[\lambda_2 P_1(t)^{(2)} - \mu_2 P_2(t)^{(2)}] + P_2(t)^{(2)}\,[-\lambda_1 P_1(t)^{(1)} + \mu_1 P_2(t)^{(1)}]$$

$$= \lambda_2\, P_1(t)^{(1)}\, P_1(t)^{(2)} - \lambda_2\, P_1(t)^{(1)}\, P_2(t)^{(2)} - \lambda_1 P_1(t)^{(1)}\, P_2(t)^{(2)} + \mu_1 P_2(t)^{(1)}\, P_2(t)^{(2)}$$

$$=\lambda_2\, Q_1(t) - (\mu_2 + \mu_1)\, Q_2(t) + \mu_1 Q_4(t)$$

$$\frac{dQ_3(t)}{dt} = P_2(t)^{(1)}\, dP_1(t)^{(2)}/dt + P_1(t)^{(2)}\, dP_2(t)^{(1)}/dt$$

$$=P_2(t)^{(1)}\,[-\lambda_2 P_1(t)^{(2)} + \mu_2 P_2(t)^{(2)}] + P_1(t)^{(2)}\,[\lambda_1 P_1(t)^{(1)} - \mu_1 P_2(t)^{(1)}]$$

$$= -\lambda_2\, P_2(t)^{(1)}\, P_1(t)^{(2)} + \mu_2\, P_2(t)^{(1)}\, P_2(t)^{(2)}$$

$$+\lambda_1 P_1(t)^{(1)}\, P_1(t)^{(2)} - \mu_1 P_2(t)^{(1)}\, P_1(t)^{(2)}$$

$$= -\mu_2\, Q_3(t) + \mu_2\, Q_4(t) + \lambda_1 Q_1(t) - \mu_1 Q_3(t)$$

$$= \lambda_1 Q_1(t) - (\lambda_2 + \mu_1) Q_3(t) + \mu_2\, Q_4(t)$$

$$\frac{dQ_4(t)}{dt} = P_2(t)^{(1)}\, dP_2(t)^{(2)}/dt + P_2(t)^{(2)}\, dP_2(t)^{(1)}/dt$$

$$=P_2(t)^{(1)}\,[\lambda_2 P_1(t)^{(2)} - \lambda_2 P_2(t)^{(2)}] + P_2(t)^{(2)}\,[\lambda_1 P_1(t)^{(1)} - \mu_1 P_2(t)^{(1)}]$$

$$= \lambda_2\, P_2(t)^{(1)}\, P_1(t)^{(2)} - \mu_2\, P_2(t)^{(1)}\, P_2(t)^{(2)}$$

$$+\lambda_1 P_1(t)^{(1)}\, P_2(t)^{(2)} - \lambda_1 P_2(t)^{(1)}\, P_2(t)^{(2)}$$

$$= \lambda_2\, Q_3(t) - \lambda_2\, Q_4(t) + \lambda_1 Q_2(t) - \mu_1 Q_4(t)$$

$$= \lambda_1 Q_2\,(t) + \lambda_2\, Q_3(t) - (\mu_1 + \mu_2) Q_4(t)$$

The complete system differential equation can be written as:

$$\frac{dQ_1(t)}{dt} = -(\lambda_1 + \mu_2)Q_1(t) + \mu_2\, Q_2(t) + \mu_1 Q_3(t)$$

$$\frac{dQ_2(t)}{dt} = \lambda_2\, Q_1(t) - (\lambda_2 + \lambda_1)Q_2(t) + \mu_1 Q_4(t)$$

$$\frac{dQ_3(t)}{dt} = \lambda_1 Q_1(t) - (\lambda_2 + \mu_1)Q_3\,(t) + \mu_2\, Q_4(t)$$

$$\frac{dQ_4(t)}{dt} = \lambda_1 Q_2\,(t) + \lambda_2\, Q_3(t) - (\mu_1 + \mu_2)\, Q_4(t)$$

$$\begin{bmatrix} \frac{dQ_1(t)}{dt} \\ \frac{dQ_2(t)}{dt} \\ \frac{dQ_3(t)}{dt} \\ \frac{dQ_4(t)}{dt} \end{bmatrix} = \begin{bmatrix} -(\lambda_1+\lambda_2) & \mu_2 & \mu_1 & 0 \\ \lambda_2 & -(\mu_2+\lambda_1) & 0 & \mu_1 \\ \lambda_1 & 0 & -(\lambda_2+\mu_1) & \mu_2 \\ 0 & \lambda_1 & \lambda_2 & -(\mu_1+\mu_2) \end{bmatrix} \begin{bmatrix} Q_1(t) \\ Q_2(t) \\ Q_3(t) \\ Q_4(t) \end{bmatrix}$$

The state space-diagram for the above equation is described in Figure 7.3.

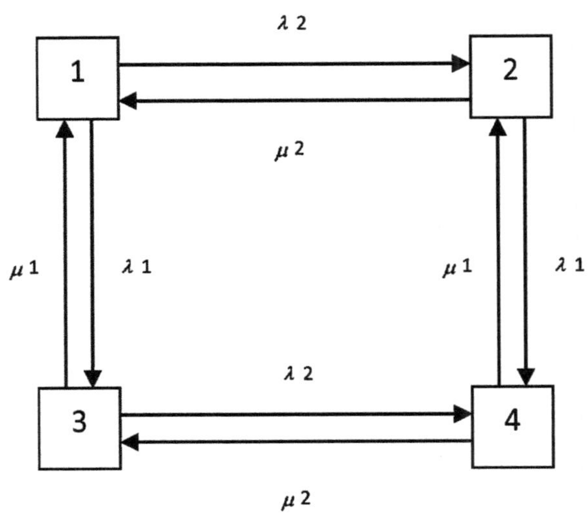

FIGURE 7.3 The overall system state-space diagram of the two subsystems.

The overall system characterized by four states because each system can be in either of two states, which are compound from the simple two-states for each sub-system. The above method can be readily being generalized to a system containing N sub-systems. Again assuming that each sub-system is one of two states, one sees that the overall system can be described by the following two states given by:

$$\left.\begin{array}{l}
Q_1(t) = P_1(t)^{(1)}P_1(t)^{(2)}P_1(t)^{(3)}\ldots\ldots\ldots\ldots P_1(t)^{(N-2)}P_1(t)^{(N-1)}P_1(t)^{(N)} \\
Q_2(t) = P_1(t)^{(1)}P_1(t)^{(2)}P_1(t)^{(3)}\ldots\ldots\ldots\ldots P_1(t)^{(N-2)}P_1(t)^{(N-1)}P_2(t)^{(N)} \\
Q_3(t) = P_1(t)^{(1)}P_1(t)^{(2)}P_1(t)^{(3)}\ldots\ldots\ldots\ldots P_1(t)^{(N-2)}P_2(t)^{(N-1)}P_1(t)^{(N)} \\
Q_4(t) = P_1(t)^{(1)}P_1(t)^{(2)}P_1(t)^{(3)}\ldots\ldots\ldots\ldots P_2(t)^{(N-2)}P_1(t)^{(N-1)}P_1(t)^{(N)} \\
\quad\cdot \qquad \cdot \qquad \cdot \qquad \cdot \qquad \cdot \qquad \cdot \\
\quad\cdot \qquad \cdot \qquad \cdot \qquad \cdot \qquad \cdot \qquad \cdot \\
\quad\cdot \qquad \cdot \qquad \cdot \qquad \cdot \qquad \cdot \qquad \cdot \\
Q_{M-3}(t) = P_2(t)^{(1)}P_2(t)^{(2)}P_2(t)^{(3)}\ldots\ldots\ldots\ldots P_1(t)^{(N-2)}P_2(t)^{(N-1)}P_2(t)^{(N)} \\
Q_{M-2}(t) = P_2(t)^{(1)}P_2(t)^{(2)}P_2(t)^{(3)}\ldots\ldots\ldots\ldots P_2(t)^{(N-2)}P_1(t)^{(N-1)}P_2(t)^{(N)} \\
Q_{M-1}(t) = P_2(t)^{(1)}P_2(t)^{(2)}P_2(t)^{(3)}\ldots\ldots\ldots\ldots P_2(t)^{(N-2)}P_2(t)^{(N-1)}P_1(t)^{(N)} \\
Q_M(t) = P_2(t)^{(1)}P_2(t)^{(2)}P_2(t)^{(3)}\ldots\ldots\ldots\ldots P_2(t)^{(N-2)}P_2(t)^{(N-1)}P_2(t)^{(N)}
\end{array}\right\}$$

One has $M = (2)^N$, where M is the total number of the equivalent system states of N.

7.6 APPLICATIONS

As stated earlier, the Kronecker product (Shakeri et al., 2016; Mariet et al., 2016; Kepner et al., 2018) is easier and faster the differentiation technique in forming the transient rate matrix. Let λ_1=.1, λ_2=.2, λ_3=.3, λ_4=.4 and λ_5=0.5, and let μ_1=0.6, μ_2=0.7, μ_3=0.8, μ_4=0.9 and μ_5=0. The parameters were chosen to simplify the calculations. So, to build a system from two subsystems let $\lambda = [0.1\ 0.2]$ and $\mu = [0.60.7]$ using the differentiation technique following the equations from (7.10) to (7.12) will give the following:

$$dP_1(t)^{(1)}/dt = -\lambda_1 P_1(t)^{(1)} + \mu_1 P_2(t)^{(1)}$$
$$dP_2(t)^{(1)}/dt = \lambda_1 P_1(t)^{(1)} - \mu_1 P_2(t)^{(1)}$$

$$dP_1(t)^{(2)}/dt = -\lambda_2 P_1(t)^{(2)} + \mu_2 P_2(t)^{(2)}$$
$$dP_2(t)^{(2)}/dt = \lambda_2 P_1(t)^{(2)} - \mu_2 P_2(t)^{(2)}$$

$$dP_1(t)^{(1)}/dt = -.1P_1(t)^{(1)} + .6P_2(t)^{(1)}$$
$$dP_2(t)^{(1)}/dt = .1P_1(t)^{(1)} - .6P_2(t)^{(1)}$$
$$dP_1(t)^{(2)}/dt = -.2P_1(t)^{(2)} + .7P_2(t)^{(2)}$$
$$dP_2(t)^{(2)}/dt = .2P_1(t)^{(2)} - .7P_2(t)^{(2)}$$

This will lead to the following matrix

$$\begin{bmatrix} \frac{dQ_1(t)}{dt} \\ \frac{dQ_2(t)}{dt} \\ \frac{dQ_3(t)}{dt} \\ \frac{dQ_4(t)}{dt} \end{bmatrix} = \begin{bmatrix} -(.1+.2) & .7 & .6 & 0 \\ .2 & -(.7+.1) & 0 & .6 \\ .1 & 0 & -(.2+.6) & .7 \\ 0 & .1 & .2 & -(.6+.7) \end{bmatrix} \begin{bmatrix} Q_1(t) \\ Q_2(t) \\ Q_3(t) \\ Q_4(t) \end{bmatrix}$$

$$\begin{bmatrix} \frac{dQ_1(t)}{dt} \\ \frac{dQ_2(t)}{dt} \\ \frac{dQ_3(t)}{dt} \\ \frac{dQ_4(t)}{dt} \end{bmatrix} = \begin{bmatrix} -.3 & .7 & .6 & 0 \\ .2 & -.8 & 0 & .6 \\ .1 & 0 & -.8 & .7 \\ 0 & .1 & .2 & -1.3 \end{bmatrix} \begin{bmatrix} Q_1(t) \\ Q_2(t) \\ Q_3(t) \\ Q_4(t) \end{bmatrix}$$

Using Kronecker product is much simpler than the differentiation technique. The first step to using the transition rates between the states that the system is passing through to form the matrix that represents the systems. The second step is feeding the matrices to the Kronecker Sum program.

$$\text{Unit one} = \begin{bmatrix} -.1 & .6 \\ .1 & -.6 \end{bmatrix}$$

$$\text{Unit two} = \begin{bmatrix} -.2 & .7 \\ .2 & -.7 \end{bmatrix}$$

$$\text{First Unit} \oplus \text{Second Unit} = \begin{bmatrix} -.3 & .7 & .6 & 0 \\ .2 & -.8 & 0 & .6 \\ .1 & 0 & -.8 & .7 \\ 0 & .1 & .2 & -1.3 \end{bmatrix}$$

From the above, it can be seen that the use of the Kronecker product is much simpler and faster than the differentiation technique. Following is the sum of three, four, and five subsystems using the same technique.

$$\text{Unit one} = \begin{bmatrix} -.1 & .6 \\ .1 & -.6 \end{bmatrix}, \text{Unit two} = \begin{bmatrix} -.2 & .7 \\ .2 & -.7 \end{bmatrix}, \text{Unit three} = \begin{bmatrix} -.3 & .8 \\ .3 & -.8 \end{bmatrix},$$

$$\text{Unit four} = \begin{bmatrix} -.4 & .9 \\ .4 & -.9 \end{bmatrix} \text{ and Unit five} = \begin{bmatrix} -.5 & 0 \\ .5 & 0 \end{bmatrix}$$

The overall matrix for three subsystems is the following

Unit one ⊕ Unit two ⊕ Unit three =

$$\begin{bmatrix}
-0.6 & 0.8 & 0.7 & 0 & 0.6 & 0 & 0 & 0 \\
0.3 & -1.1 & 0 & 0.7 & 0 & 0.6 & 0 & 0 \\
0.2 & 0 & -1.1 & 0.8 & 0 & 0 & 0.6 & 0 \\
0 & 0.2 & 0.3 & -1.6 & 0 & 0 & 0 & 0.6 \\
0.1 & 0 & 0 & 0 & -1.1 & 0.8 & 0.7 & 0 \\
0 & 0.1 & 0 & 0 & 0.3 & -1.6 & 0 & 0.7 \\
0 & 0 & 0.1 & 0 & 0.2 & 0 & -1.6 & 0.8 \\
0 & 0 & 0 & 0.1 & 0 & 0.2 & 0.3 & -2.1
\end{bmatrix}$$

Four Units = First Unit ⊕ Second Unit ⊕ Third Unit ⊕ Fourth Unit =

$$\begin{bmatrix}
-1 & 0.9 & 0.8 & 0 & 0.7 & 0 & 0 & 0 & 0.6 & 0 & 0 & 0 & 0 & 0 & 0 & 0 \\
0.4 & -1.5 & 0 & 0.8 & 0 & 0.7 & 0 & 0 & 0 & 0.6 & 0 & 0 & 0 & 0 & 0 & 0 \\
0.3 & 0 & -1.5 & 0.9 & 0 & 0 & 0.7 & 0 & 0 & 0 & 0.6 & 0 & 0 & 0 & 0 & 0 \\
0 & 0.3 & 0.4 & -2 & 0 & 0 & 0 & 0.7 & 0 & 0 & 0 & 0.6 & 0 & 0 & 0 & 0 \\
0.2 & 0 & 0 & 0 & -1.5 & 0.9 & 0.8 & 0 & 0 & 0 & 0 & 0 & 0.6 & 0 & 0 & 0 \\
0 & 0.2 & 0 & 0 & 0.4 & -2 & 0 & 0.8 & 0 & 0 & 0 & 0 & 0 & 0.6 & 0 & 0 \\
0 & 0 & 0.2 & 0 & 0.3 & 0 & -2 & 0.9 & 0 & 0 & 0 & 0 & 0 & 0 & 0.6 & 0 \\
0 & 0 & 0 & 0.2 & 0 & 0.3 & 0.4 & -2.5 & 0 & 0 & 0 & 0 & 0 & 0 & 0 & 0.6 \\
0.1 & 0 & 0 & 0 & 0 & 0 & 0 & 0 & -1.5 & 0.9 & 0.8 & 0 & 0.7 & 0 & 0 & 0 \\
0 & 0.1 & 0 & 0 & 0 & 0 & 0 & 0 & 0.4 & -2 & 0 & 0.8 & 0 & 0.7 & 0 & 0 \\
0 & 0 & 0.1 & 0 & 0 & 0 & 0 & 0 & 0.3 & 0 & -2 & 0.9 & 0 & 0 & 0.7 & 0 \\
0 & 0 & 0 & 0.1 & 0 & 0 & 0 & 0 & 0 & 0.3 & 0.4 & -2.5 & 0 & 0 & 0 & 0.7 \\
0 & 0 & 0 & 0 & 0.1 & 0 & 0 & 0 & 0.2 & 0 & 0 & 0 & -2 & 0.9 & 0.8 & 0 \\
0 & 0 & 0 & 0 & 0 & 0.1 & 0 & 0 & 0 & 0.2 & 0 & 0 & 0.4 & -2.5 & 0 & 0.8 \\
0 & 0 & 0 & 0 & 0 & 0 & 0.1 & 0 & 0 & 0 & 0.2 & 0 & 0.3 & 0 & -2.5 & 0.9 \\
0 & 0 & 0 & 0 & 0 & 0 & 0 & 0.1 & 0 & 0 & 0 & 0.2 & 0 & 0.3 & 0.4 & -3
\end{bmatrix}$$

Five Units = Unit one ⊕ Unit two ⊕ Unit three ⊕ Unit four ⊕ Unit five =

$$
\begin{bmatrix}
-1.5 & 0 & 0.9 & 0 & 0.8 & 0 & 0 & 0 & 0.7 & 0 & 0 & 0 & 0 & 0 & 0 & 0 & 0.6 & 0 & 0 & 0 & 0 & 0 & 0 & 0 & 0 & 0 & 0 & 0 & 0 & 0 & 0 & 0 \\
0.5 & -1 & 0 & 0.9 & 0 & 0.8 & 0 & 0 & 0 & 0.7 & 0 & 0 & 0 & 0 & 0 & 0 & 0 & 0.6 & 0 & 0 & 0 & 0 & 0 & 0 & 0 & 0 & 0 & 0 & 0 & 0 & 0 & 0 \\
0.4 & 0 & -2 & 0 & 0 & 0 & 0.8 & 0 & 0 & 0 & 0.7 & 0 & 0 & 0 & 0 & 0 & 0 & 0 & 0.6 & 0 & 0 & 0 & 0 & 0 & 0 & 0 & 0 & 0 & 0 & 0 & 0 & 0 \\
0 & 0.4 & 0.5 & -1.5 & 0 & 0 & 0 & 0.8 & 0 & 0 & 0 & 0.7 & 0 & 0 & 0 & 0 & 0 & 0 & 0 & 0.6 & 0 & 0 & 0 & 0 & 0 & 0 & 0 & 0 & 0 & 0 & 0 & 0 \\
0.3 & 0 & 0 & 0 & -2 & 0 & 0.9 & 0 & 0 & 0 & 0 & 0 & 0.7 & 0 & 0 & 0 & 0 & 0 & 0 & 0 & 0.6 & 0 & 0 & 0 & 0 & 0 & 0 & 0 & 0 & 0 & 0 & 0 \\
0 & 0.3 & 0 & 0 & 0.5 & -1.5 & 0 & 0.9 & 0 & 0 & 0 & 0 & 0 & 0.7 & 0 & 0 & 0 & 0 & 0 & 0 & 0 & 0.6 & 0 & 0 & 0 & 0 & 0 & 0 & 0 & 0 & 0 & 0 \\
0 & 0 & 0.3 & 0 & 0.4 & 0 & -2.5 & 0 & 0 & 0 & 0 & 0 & 0 & 0 & 0.7 & 0 & 0 & 0 & 0 & 0 & 0 & 0 & 0.6 & 0 & 0 & 0 & 0 & 0 & 0 & 0 & 0 & 0 \\
0 & 0 & 0 & 0.3 & 0 & 0.4 & 0.5 & -2 & 0 & 0 & 0 & 0 & 0 & 0 & 0 & 0.7 & 0 & 0 & 0 & 0 & 0 & 0 & 0 & 0.6 & 0 & 0 & 0 & 0 & 0 & 0 & 0 & 0 \\
0.2 & 0 & 0 & 0 & 0 & 0 & 0 & 0 & -2 & 0 & 0.9 & 0 & 0.8 & 0 & 0 & 0 & 0 & 0 & 0 & 0 & 0 & 0 & 0 & 0 & 0.6 & 0 & 0 & 0 & 0 & 0 & 0 & 0 \\
0 & 0.2 & 0 & 0 & 0 & 0 & 0 & 0 & 0.5 & -1.5 & 0 & 0.9 & 0 & 0.8 & 0 & 0 & 0 & 0 & 0 & 0 & 0 & 0 & 0 & 0 & 0 & 0.6 & 0 & 0 & 0 & 0 & 0 & 0 \\
0 & 0 & 0.2 & 0 & 0 & 0 & 0 & 0 & 0.4 & 0 & -2.5 & 0 & 0 & 0 & 0.8 & 0 & 0 & 0 & 0 & 0 & 0 & 0 & 0 & 0 & 0 & 0 & 0.6 & 0 & 0 & 0 & 0 & 0 \\
0 & 0 & 0 & 0.2 & 0 & 0 & 0 & 0 & 0 & 0.4 & 0.5 & -2 & 0 & 0 & 0 & 0.8 & 0 & 0 & 0 & 0 & 0 & 0 & 0 & 0 & 0 & 0 & 0 & 0.6 & 0 & 0 & 0 & 0 \\
0 & 0 & 0 & 0 & 0.2 & 0 & 0 & 0 & 0.3 & 0 & 0 & 0 & -2.5 & 0 & 0.9 & 0 & 0 & 0 & 0 & 0 & 0 & 0 & 0 & 0 & 0 & 0 & 0 & 0 & 0.6 & 0 & 0 & 0 \\
0 & 0 & 0 & 0 & 0 & 0.2 & 0 & 0 & 0 & 0.3 & 0 & 0 & 0.5 & -2 & 0 & 0.9 & 0 & 0 & 0 & 0 & 0 & 0 & 0 & 0 & 0 & 0 & 0 & 0 & 0 & 0.6 & 0 & 0 \\
0 & 0 & 0 & 0 & 0 & 0 & 0.2 & 0 & 0 & 0 & 0.3 & 0 & 0.4 & 0 & -3 & 0 & 0 & 0 & 0 & 0 & 0 & 0 & 0 & 0 & 0 & 0 & 0 & 0 & 0 & 0 & 0.6 & 0 \\
0 & 0 & 0 & 0 & 0 & 0 & 0 & 0.2 & 0 & 0 & 0 & 0.3 & 0 & 0.4 & 0.5 & -2.5 & 0 & 0 & 0 & 0 & 0 & 0 & 0 & 0 & 0 & 0 & 0 & 0 & 0 & 0 & 0 & 0.6 \\
0.1 & 0 & 0 & 0 & 0 & 0 & 0 & 0 & 0 & 0 & 0 & 0 & 0 & 0 & 0 & 0 & -2 & 0 & 0.9 & 0 & 0.8 & 0 & 0 & 0 & 0.7 & 0 & 0 & 0 & 0 & 0 & 0 & 0 \\
0 & 0.1 & 0 & 0 & 0 & 0 & 0 & 0 & 0 & 0 & 0 & 0 & 0 & 0 & 0 & 0 & 0.5 & -1.5 & 0 & 0.9 & 0 & 0.8 & 0 & 0 & 0 & 0.7 & 0 & 0 & 0 & 0 & 0 & 0 \\
0 & 0 & 0.1 & 0 & 0 & 0 & 0 & 0 & 0 & 0 & 0 & 0 & 0 & 0 & 0 & 0 & 0.4 & 0 & -2.5 & 0 & 0 & 0 & 0.8 & 0 & 0 & 0 & 0.7 & 0 & 0 & 0 & 0 & 0 \\
0 & 0 & 0 & 0.1 & 0 & 0 & 0 & 0 & 0 & 0 & 0 & 0 & 0 & 0 & 0 & 0 & 0 & 0.4 & 0.5 & -2 & 0 & 0 & 0 & 0.8 & 0 & 0 & 0 & 0.7 & 0 & 0 & 0 & 0 \\
0 & 0 & 0 & 0 & 0.1 & 0 & 0 & 0 & 0 & 0 & 0 & 0 & 0 & 0 & 0 & 0 & 0.3 & 0 & 0 & 0 & -2.5 & 0 & 0.9 & 0 & 0 & 0 & 0 & 0 & 0.7 & 0 & 0 & 0 \\
0 & 0 & 0 & 0 & 0 & 0.1 & 0 & 0 & 0 & 0 & 0 & 0 & 0 & 0 & 0 & 0 & 0 & 0.3 & 0 & 0 & 0.5 & -2 & 0 & 0.9 & 0 & 0 & 0 & 0 & 0 & 0.7 & 0 & 0 \\
0 & 0 & 0 & 0 & 0 & 0 & 0.1 & 0 & 0 & 0 & 0 & 0 & 0 & 0 & 0 & 0 & 0 & 0 & 0.3 & 0 & 0.4 & 0 & -3 & 0 & 0 & 0 & 0 & 0 & 0 & 0 & 0.7 & 0 \\
0 & 0 & 0 & 0 & 0 & 0 & 0 & 0.1 & 0 & 0 & 0 & 0 & 0 & 0 & 0 & 0 & 0 & 0 & 0 & 0.3 & 0 & 0.4 & 0.5 & -2.5 & 0 & 0 & 0 & 0 & 0 & 0 & 0 & 0.7 \\
0 & 0 & 0 & 0 & 0 & 0 & 0 & 0 & 0.1 & 0 & 0 & 0 & 0 & 0 & 0 & 0 & 0.2 & 0 & 0 & 0 & 0 & 0 & 0 & 0 & -2.5 & 0 & 0.9 & 0 & 0.8 & 0 & 0 & 0 \\
0 & 0 & 0 & 0 & 0 & 0 & 0 & 0 & 0 & 0.1 & 0 & 0 & 0 & 0 & 0 & 0 & 0 & 0.2 & 0 & 0 & 0 & 0 & 0 & 0 & 0.5 & -2 & 0 & 0.9 & 0 & 0.8 & 0 & 0 \\
0 & 0 & 0 & 0 & 0 & 0 & 0 & 0 & 0 & 0 & 0.1 & 0 & 0 & 0 & 0 & 0 & 0 & 0 & 0.2 & 0 & 0 & 0 & 0 & 0 & 0.4 & 0 & -3 & 0 & 0 & 0 & 0.8 & 0 \\
0 & 0 & 0 & 0 & 0 & 0 & 0 & 0 & 0 & 0 & 0 & 0.1 & 0 & 0 & 0 & 0 & 0 & 0 & 0 & 0.2 & 0 & 0 & 0 & 0 & 0 & 0.4 & 0.5 & -2.5 & 0 & 0 & 0 & 0.8 \\
0 & 0 & 0 & 0 & 0 & 0 & 0 & 0 & 0 & 0 & 0 & 0 & 0.1 & 0 & 0 & 0 & 0 & 0 & 0 & 0 & 0.2 & 0 & 0 & 0 & 0.3 & 0 & 0 & 0 & -3 & 0 & 0.9 & 0 \\
0 & 0 & 0 & 0 & 0 & 0 & 0 & 0 & 0 & 0 & 0 & 0 & 0 & 0.1 & 0 & 0 & 0 & 0 & 0 & 0 & 0 & 0.2 & 0 & 0 & 0 & 0.3 & 0 & 0 & 0.5 & -2.5 & 0 & 0.9 \\
0 & 0 & 0 & 0 & 0 & 0 & 0 & 0 & 0 & 0 & 0 & 0 & 0 & 0 & 0.1 & 0 & 0 & 0 & 0 & 0 & 0 & 0 & 0.2 & 0 & 0 & 0 & 0.3 & 0 & 0.4 & 0 & -3.5 & 0 \\
0 & 0 & 0 & 0 & 0 & 0 & 0 & 0 & 0 & 0 & 0 & 0 & 0 & 0 & 0 & 0.1 & 0 & 0 & 0 & 0 & 0 & 0 & 0 & 0.2 & 0 & 0 & 0 & 0.3 & 0 & 0.4 & 0.5 & -3
\end{bmatrix}
$$

7.7 NUMBER OF CONNECTED SUBSYSTEM

In this section, two, three, and four identical power generators connected to find out the steady-state probabilities of the connected generators to be connected (Shakeri et al., 2016; Mariet et al., 2016; Kepner et al., 2018). The generators have the same λ =.1 and μ =.9. Table 7.5 shows the number of generators connected and its respective P_1. Figure 7.4 represents the P_1 with a respective number of generators. On the same hand, applying the curve-fitting technique to find the most suitable formula for the relationship between the numbers of the generating-unit versus the first steady-state probability. The formula is becoming:

$$y = 0.9995 - 0.1047167*x + 0.00525*x^2 - 0.0001333333*x^3 + 3.972683*10^{-18}*x^4$$

TABLE 7.5 Steady-State Probability for Different Number of Generators

Number of Generators	First Steady-State Probability
2	0.8100
3	0.7290
4	0.6561
5	0.5905
6	0.5314

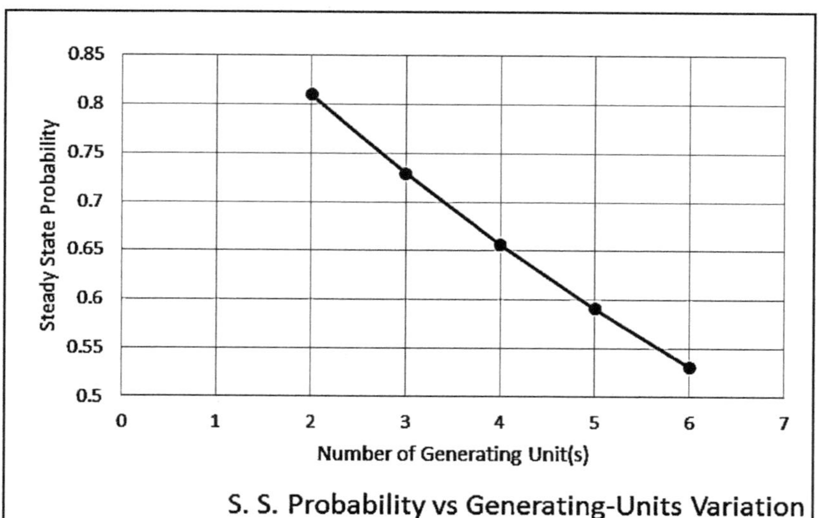

FIGURE 7.4 First steady-state probability versus number of generators.

From Table 7.5 and Figure 7.4, it easy to see that a minimum number of connected generators illustrate the system with a lower number of generators is more reliable than that with a higher number of generating units.

KEYWORDS

- **Kronecker product modeling**
- **Kronecker products**
- **Kronecker sum**
- **Kronecker technique**
- **matrix sum**
- **system building**

REFERENCES

Broxson, B. J. "The Kronecker Product," University of North Florida (UNF) digital commons. *A Master Thesis*, Submitted in Mathematics, Spring, **2006**.

Kepner, J., Samsi, S., Arcand, W., Bestor, D., Bergeron, B., Davis, T., et al. Design, generation, and validation of extreme scale power-law graphs. *IEEE IPDPS 2018 Graph Algorithm Building Blocks (GABB) Workshop, USA*, **2018**. doi: 10.1109/IPDPSW.2018.00055.

Mariet, Z., & Suvrit, S. Kronecker determinantal point processes, machine learning (cs.LG), artificial intelligence (cs.AI), machine learning (stat.ML). *MIT*, **2016**, arXiv:1605.08374.

Shakeri, Z., Bajwa, W. U., & Sarwate, A. D. Minimax lower bounds for Kronecker-structured dictionary learning, conference: 2016. *IEEE International Symposium on Information Theory (ISIT), USA*, **2016**. doi: 10.1109/ISIT.2016.7541479.

Society for Industrial and Applied Mathematics, **2005**. Retrieved from: http://www.siam.org/books/textbooks/OT91sample.pdf (Accessed on 12 October 2019).

Van Loan, C. F. The ubiquitous Kronecker product. *Journal of Computational and Applied Mathematics*, **2000**, *123*, 85–100.

CHAPTER 8

Transient Probabilities Solution

8.1 INTRODUCTION

Any system under study needs a transient study status. For this reason, the problem of obtaining transient probabilities solution addressed in the present chapter. Going through this type of study is helping to find not only the transient solution, but also the steady-state probabilities for the overall electric power system. The transient solution is necessary, where the importance came from the role of the transient solution in studying the behavior of power system model immediately after start-up or repairs the system, or even the behavior is needed over a certain limit period through which the system is required to perform a given mission (Thomson, 1018).

To obtain the transient solution, there are many methods to find the solution of the electric power system discussed in the literature. However, in the present chapter, three different methods are discussed and studied. These three are:

- Fourth order Runge-Kutta method.
- System multiplication method (SMM).
- Adams method (using Matlab Simulink).

The aim of this chapter is to show how transient probabilities for any model can be conveniently obtained numerically.

8.2 FOURTH-ORDER RUNGE-KUTTA METHOD

The German mathematicians Carl Runge (1856–1927) and Wilhelm Kutta (1867–1944) the Runge-Kutta method was introduced the fourth-order method, which is widely used in much present-day computer application

(Thomson, 2018). The reason behind that is the computation speed, precision, and relatively large interval of stability (Tang, 2018). The algorithms for the method popularly known simply as Rung-Kutta method which can be written as follows:

$$p_{i+1}^{j} = p_1^{i} + hf_j(t_i, p_i^{(1)}, p_i^{(2)}, \ldots\ldots, p_i^{(N)}, h)$$

$$f_j(t_i, p_i^{(1)}, p_i^{(2)}, \ldots\ldots, p_i^{(N)}, h) = \frac{1}{6}(k_1 p_j + k_2 p_j + k_3 p_j + k_4 p_j)$$

$$k_1 p_j = f_j(t_i, p_1, p_2, \ldots\ldots, p_N)$$

$$k_2 p_j = f_j(t_i + \frac{1}{2}h, p_{1i} + \frac{1}{2}hk_1 p_1, p_{2i} + \frac{1}{2}hk_1 p_2, \ldots\ldots, p_{Ni} + \frac{1}{2}hk_1 p_N)$$

$$k_3 p_j = f_j(t_i + \frac{1}{2}h, p_{1i} + \frac{1}{2}hk_2 p_1, p_{2i} + \frac{1}{2}hk_2 p_2, \ldots\ldots, p_{Ni} + \frac{1}{2}hk_2 p_N)$$

$$k_4 p_j = f_j(t_i + \frac{1}{2}h, p_{1i} + \frac{1}{2}hk_3 p_1, p_{2i} + \frac{1}{2}hk_3 p_2, \ldots\ldots, p_{Ni} + \frac{1}{2}hk_3 p_N)$$

where: *i* is the *i* iteration number (i = 1 to n).; *j* is the function number (j = 1 to N).; *i* and *j* are integers.

$t_i = t_0 + ih$

In Tang's (2018) study, the author developed two ways to construct Runge-Kutta type methods of a randomly high order. In the construction of Runge-Kutta type methods, a critical technique associated with the orthogonal polynomial expansion is applied. By using this approach, we do not need to study the simple solution of multivariable nonlinear algebraic equations restricting from order conditions. The author (Tang, 2018) provides a note on continuous-stage Runge-Kutta methods (csRK) for solving initial value problems of first-order ordinary differential equations. These methods are an interesting and creative extension of traditional Runge-Kutta methods, it can give us a new perspective on Runge-Kutta discretization, and it may enlarge the application of its approximation theory in modern mathematics and engineering fields. A highlighted advantage of the investigation of csRK methods that it does not need to study the tedious solution of multivariable nonlinear algebraic equations associated with order conditions. At the same time, the author review, discuss, and further promote the recently-developed csRK theory. In the construction of RK-type methods, a crucial technique associated with the orthogonal polynomial expansion is fully utilized. By using this approach, we do not need to study the tedious solution of multivariable nonlinear algebraic equations stemming from order conditions.

The classical continuous Runge-Kutta methods are widely applied to compute the numerical solutions of delay differential equations without impulsive perturbations. However, the classical continuous Runge-Kutta methods cannot be applied directly to impulsive delay differential equations, because the exact solutions of the impulsive delay differential equations are not continuous. In Zhang and Song (2019) study they impulsive continuous Runge-Kutta methods which are constructed for impulsive delay differential equations with the variable delay based on the theory of continuous Runge-Kutta methods, convergence of the constructed numerical methods is studied, and some numerical examples are given to confirm the theoretical results (Christopher et al., 2019).

Two new implicit-explicit additive Runge-Kutta (ARK2) methods are given with fourth-and fifth-order formal accuracies, respectively. Both methods are in (Kennedya, Mark, and Carpenterb, 2019). Both methods are combining explicit Runge-Kutta (ERK) methods with explicit singly-diagonally implicit Runge-Kutta (ESDIRK) methods and include an embedded method for error control. The two methods have ESDIRKs that are internally L-stable on stages three and higher, have only modestly negative Eigenvalues to the stage and step algebraic-stability matrices, and have stage-order two. To improve computational efficiency, the fourth-order method has a diagonal coefficient of 0.1235. This is concluding to offset much of the extra computational cost of an extra stage by facilitating iterative convergence at each stage. Linear stability domains for both ERK methods have been made quite large, and the dominant coupling stability term between the stability of the ESDIRK and ERK for very stiff modes has been removed. As well, the fourth-order method is one of the best all-around fourth-order IMEX ARK2, which the authors are aware of. The fifth-order method is likely best suited to mildly stiff problems with tight error tolerances. The methods were tested. The obtained results suggested that these new methods represent an improvement over existing methods of the same class.

8.3 SYSTEM MULTIPLICATION METHOD (SMM)

In many existed practical problems, a natural unit of time often suggested itself, i.e., if a person modeling the behavior of weather, a convenient unit of time could be a week. This natural discrimination of time also often has an advantage in numerical evaluation (Thomson, 2018). The transformation from the continuous-time to the discrete-time domain can be accomplished as follows by considering:

$$\frac{dP(t)}{dt} = AP(t)$$

where A is the transition rate matrix. One can approximate the differentiation of the probability as a function of time to the rate of time as:

$$\frac{dP(t)}{dt} \text{ becomes as } \frac{\{P(t+h) - P(t)\}}{h}$$

The variable h is a sufficiently small interval of time. To study any system, the researcher assumed the variable (h) as an interval. Nevertheless, this gives:

$$P(t+h) - P(t) = h.AP(t)$$

i.e.,

$$P(t+h) = P(t) + h.AP(t)$$

$$P(t+h) = (I + h.A).P(t)$$

$$P(t+h) = T.P(t)$$

where T is called a transition matrix defined now by:

$$T = (I + h.A)$$

and I is the unit matrix.

Therefore, the algorithm for the present method can be written as follows:

$$P_{i+1} = hAP_i$$

8.4 ADAMS METHOD

The technique called the Adams method (Ameen, and Novati, 2017) is used to find the numerical solutions of initial value problems. The effectiveness of this technique is helping for the treatment of problems based on its attractive properties and an efficient technique. It deals with the algebraic nonlinear systems. The Adams method is a predictor-corrector and multi-step method. The principle behind using this method is the use of the past calculated values of P(t) to build a polynomial that approximates the derivative function and extrapolate this into the next interval (Yanga and Ralescu, 2015).

The general formula of the Adams method is as follows:

$$(P_{i+1})_p = P_i + \frac{h}{24}(55f_i - 59f_{i-1} + 37f_{i-2} - 9f_{i-3}) + \frac{251}{720}h^5 P^V(\xi_1)$$

and

$$(P_{i+1})_c = P_i + \frac{h}{24}(9f_{i+1} + 19f_i - 5f_{i-1} + f_{i-2}) + \frac{19}{720}h^5 P^V(\xi_2)$$

where

$$t_{i-3} < \xi_1, \xi_2 < t_{i+1}$$

Elimination of error terms gives the formula:

$$P_{i+1} = \{19(P_{i+1})_P + 251(P_{i+1})_c\}/270$$

8.5 INCREMENTAL OF TIME LIMITATION (H)

The SMM, Runge-Kutta, and Adams methods are compared, and the way of obtaining solutions of prescribed accuracy is outlined for the methods which numerically integrate the differential equation. The step length used determines the accuracy of the integration methods, and so it is useful to be able to estimate the step length to give the required accuracy. In principle, the step length is determined by calculating two rough estimates of the solution and using them to determine the final step length.

For some $p \geq 1$, the error is $0(h^p)$. So, to estimate a suitable h, the following procedure is adopted:

Let x be the true value to be estimated.
Using step size h_1 to obtain an estimate value r_1.
Using step size h_2 to obtain an estimate value r_2.

Suppose the errors are e_1 and e_2, respectively, where: $e_1 = zh_1^p$

Then the estimated values are:

$$r_1 = x + e_1$$
$$r_2 = x + e_2$$

Divide step size h_1 by n (an integer) to obtain h_2.

$$h_2 = h_1/n$$

Then

$$\begin{aligned} e_2 &= n(zh_2^p) \\ &= n[z(h_1 / n)^p] \\ &= n[z(h_1^p / n^p)] \\ &= n^{(1-p)} zh_1^p \end{aligned}$$

where n is the step error.

Now $|r_1 - r_2| = |e_1 - e_2|$

$$= zh_1^p \,|1 - n^{(1-p)}|$$

Therefore, the factor z is the same for all the estimated values.

Choosing $h_3 = h_1/l$ and following the same procedure as above for h_2, it's found that:

$$\begin{aligned} & e_3 = n^{(1-p)} zh_1^p \\ & \therefore n^{(1-p)} \le e_3 / (zh_1^p) \\ & \le \varepsilon / (zh_1^p) \\ & \le \frac{\varepsilon}{h_1^p} \cdot \frac{h_1^p \,|1 - n^{(1-p)}|}{|r_1 - r_2|} \\ & \le \frac{\varepsilon \,|1 - n^{(1-p)}|}{|r_1 - r_2|} \end{aligned}$$

Then

$$l \le \left[\frac{\varepsilon \,|1 - n^{(1-p)|}}{|r_1 - r_2|} \right]^{1/(1-p)}$$

Since h_1 and l are known, h_3 will be $h_3 = h_1/l$.

8.6 SYSTEM COMBINATION

The SMM, Runge-Kutta, and Adams methods are forming an easy method to help in calculating the steady-state and transient probabilities of any electric power system model. A program for calculating the transient state probabilities was obtained, as illustrated in Figure 8.1. At the same time, the steady-state probabilities are obtained through the transient solution. The combination of the presented methods is combined with the methods presented in Chapter 7.

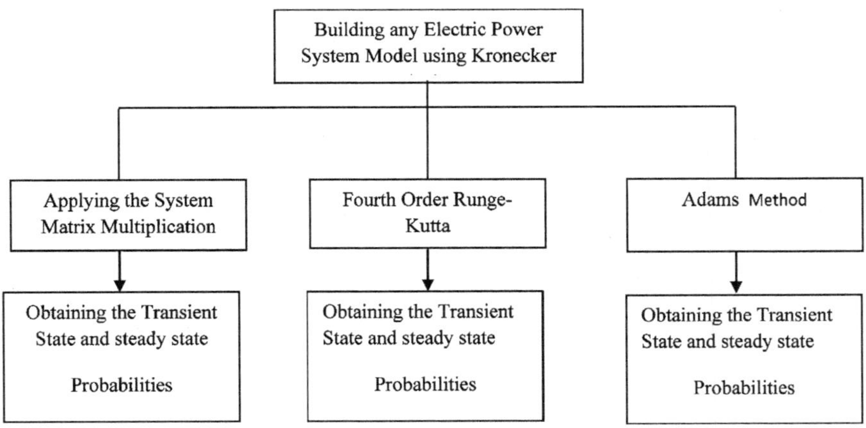

FIGURE 8.1 The proposed forecasting technique.

KEYWORDS

- **additive Runge-Kutta**
- **explicit Runge-Kutta**
- **explicit, singly-diagonally implicit Runge-Kutta**
- **system multiplication method**

REFERENCES

Ameen, I., & Novati, P. The solution of fractional order epidemic model by implicit Adams methods. *Applied Mathematical Model*, **2017,** *43*, 78–84.

Christopher, A. Kennedy, C. A., & Carpenter, M. H. Higher-order additive Runge–Kutta schemes for ordinary differential equations. *International Journal on Applied Numerical Mathematics*, **2019,** *136*, 183–205.

Kennedya, C. A., Mark, H., & Carpenterb, M. H. Higher-order additive Runge–Kutta schemes for ordinary differential equations. *Applied Numerical Mathematics*, **2019,** *136*, 183–205.

Tang, W. A Note on continuous-stage Runge-Kutta methods. *International Journal on Applied Mathematics and Computation*, **2018,** *339*, 231–241.

Thomson, W. T. *Theory of Vibration with Applications* (4th edn.). **2018,** Taylor and Francis, New York.

Yanga, X., & Ralescu, D. Adams method for solving uncertain differential equations. *Applied Mathematics and Computation*, **2015,** *270*, 993–1003.

Zhanga, G., & Song, M. Impulsive continuous Runge-Kutta methods for impulsive delay differential equations. *International Journal on Applied Mathematics and Computation*, **2019,** *341*, 160–173.

CHAPTER 9

Load Forecasting

9.1 INTRODUCTION

Accurate models for electric power load forecasting (EPLF) are essential to the operation and planning of a utility company. The load forecasting helps an electric utility to make important decisions, including decisions on purchasing and generating electric power, load switching, and infrastructure development. Load forecasts are extremely important for energy suppliers, financial institutions, and other participants in electric energy generation, transmission, distribution, and markets (He and Zheng, 2018; Jung et al., 2018).

Load forecasts can be divided into three categories (Hsu and Chen, 2003; Papadakis et al., 2003; Al-Kandari et al., 2004):

1. Short-term forecast which is usually from one hour to one week.
2. Medium forecast which is usually from a week to a year.
3. Long-term forecast which is longer than a year.

The forecasts for different time horizons are important for different operations within a utility company. The natures of these forecasts are different as well. For example, for a particular region, it is possible to predict the next day load with an accuracy of approximately 1–3%. However, it is impossible to predict the next year's peak load with a similar accuracy since accurate long-term weather forecasts are not available. For the next year's peak forecast, it is possible to provide the probability distribution of the load based on historical load data. It is also possible, according to the industry practice and economic growth to predict the next year's peak load.

Load forecasting has always been important for planning and operational decision conducted by utility companies. However, with the deregulation of the energy industries, load forecasting is even more important.

In the deregulated economy, decisions on capital expenditures based on long-term forecasting are also more important than in a non-deregulated economy when rate increases could be justified by capital expenditure projects.

Load forecasting methodologies developed may be classified into two broad categories: autonomous models and conditional models. Autonomous models attempt to relate future growth of electricity demand on a system based on its past growth, and conditional models attempt to relate it to other variables, mainly economic indicators.

9.2 LOAD FORECASTING

The load in GCC countries follows almost the same pattern, which is the peak load during the summer season. It is almost doubled the peak in the winter season. So the annual load has been divided into two categories, winter season (low season) and summer season (high season). As stated by Sayed M. Salem (Islam et al., 1995), this will minimize the error and gave a better prediction. After categorizing the load, the percentage increase between 2004 and 2005 for both categories has been calculated. Each category of the year 2005 is multiplied by its relevant percentage, and the same procedure done for the years until 2015. Each load calculated by this technique is added to the historical data and feed to the neuro-fuzzy system to predict the next year.

9.3 FUZZY LOGIC (FL) - ARTIFICIAL NEURAL NETWORK

In recent years, the artificial neural networks (ANN) and fuzzy logic (FL) systems have each providing very encouraging results in solving the problems. This has encouraging researchers to combine both ANN and FL in an attempt to create a final system that reduces the limitations of each of these individual techniques (Tamimi and Egbert, 2000). The strength of this technique lies in its ability to reduce appreciable computational time and its comparable accuracy with other modeling techniques (Metaxiotis et al., 2003) and (Padmakumari et al., 1999). The definition of a neuro-fuzzy system is a combination of ANN and fuzzy inference system (FIS) in such a way that neural network learning algorithms are used to determine the parameters of FIS (Al-Kandai et al., 2003; Mielczarski, 1995). An even more important aspect is that the system should always be interpretable in terms of fuzzy if-then rules, because it is based on the fuzzy system reflecting vague knowledge.

Neural Networks and FL are both complementary technologies in the design of intelligent systems. Each method has merits and demerits. Neural networks are essentially low-level computational structures and algorithms that offer good performance in dealing with sensory data. On the other hand, the FL techniques often deal with issues, such as reasoning, on a higher level than neural networks. However, since fuzzy systems do not have much learning capability, it is difficult for a human operator to tune the fuzzy rules and membership functions from the training data set. Also, because the internal layers of neural networks are always opaque to the user, the mapping rules in the network are not visible and are difficult to understand. Furthermore, the convergence of learning is usually very slow and not guaranteed. Thus, a promising approach for getting the benefits of both fuzzy systems and neural networks is to merge them into an integrated system. This collaboration will possess the advantages of both neural networks (e.g., learning and optimization abilities) and fuzzy systems (e.g., human-like IF-THEN rules thinking and ease of incorporating expert knowledge).

9.4 FUZZY RULES

Fuzzy reasoning usually performed using if-then rules. The fuzzy rules define the connection between input and output fuzzy (linguistic) variables. The rules consist of two parts:

- an antecedent part; and
- a consequence part.

IF Month is July and Temperature is High **THEN** Load is High

A Consequence part An Antecedent part

For more details, it can be referred in Mielczarski (1995). In the above fuzzy rules, month, temperature level and load are called fuzzy variable and July, high, and high as a linguistic variable. AND is a connective operation, OR is a union and NOT is a complement. It aggregates the results with the premise part.

9.5 FUZZY INFERENCE SYSTEM (FIS)

From a given input to output, using fuzzy logic, Fuzzy Inference (FI) is the actual process of mapping. A number of names know the FIS. These names are:

a. fuzzy model;
b. fuzzy-rule-based system;
c. simply fuzzy system;
d. fuzzy expert system;
e. fuzzy associative memory; and
f. FL controller.

Any FIS has to be defined. This FIS is needed to define the Fuzzy Knowledge Base. The Fuzzy Knowledge Base defined as a rule base having a number of fuzzy and a database that defines the membership functions of fuzzy sets.

There are a number of statements (three) coming out of the Fuzzy Knowledge Base. These three statements are Fuzzifier, Inference Engine, and Defuzzifier. The Fuzzifier is the crisp inputs are exact inputs measured by sensors and passed into the control system for processing. The Inference Engine can be defined as the engine between the Fuzzifier and Defuzzifier. The Defuzzifier of the Output Variable that represents the converging of the fuzzy output of the inference engine to a crisp using membership functions analogous to the ones used by the Fuzzifier.

9.6 NEURO-FUZZY SYSTEMS

The Electrical long-term peak load demand forecasting using a developed adaptive neuro-fuzzy inference system (ANFIS). It is important to develop a suitable model, which can be used in different types of softwares to calculate electric load forecasting and different output requested by decision-makers. The developed models should have an acceptable level of mean errors. Having a simple and pragmatic equation can provide an acceptable model to make plans for periodical operations, energy trading, and facility expansion in the electricity sector. The estimated peak load models help energy policymakers in different countries. This means that it will encourage in the development of the countries through the energy field and setting the future plan for the countries. It also helps with finding a suitable instant for electrical energy trading.

For the electricity sector decision-makers, energy trading and facility expansion are required. The electricity demand pattern is multifaceted due to the different types of loads and weather conditions. Therefore, finding an appropriate peak load-forecasting model for the country's peak demand of the electricity network is an important task for network planning and power trading. This will help in the planning of a reliable and economical operated network.

There are many exogenous variables for the long-term load forecasting (LTLF)-like weather conditions, industrial development, Population Growth, and social events in the country. For more simplicity, the model uses the main variables, which affect the peak load demand and selected for the proposed Model.

ANFIS Models are designed for output estimated Long Term Estimated Load for a number of countries. The FIS performs fuzzy reasoning. It was designed for output estimated Long Term Estimated Load for the countries if planned to estimate the peak load for the future. The MATLAB Simulink 7.10 package is suitable and well to be used to obtain the estimated peak load for the countries.

Neuro-fuzzy systems allow the incorporation of both numerical and linguistic data into the system. The neuro-fuzzy system is also capable of extracting fuzzy knowledge from numerical data. The neuro-fuzzy systems divided into two main groups, the neural FISs, and fuzzy neural networks. Various neural networks, e.g., Multilayer Perceptions (MLP) or Radial-Basis Function Networks are capable of learning nonlinear mappings and generalizing over a set of methods very accurately. Rule-based neural networks implement a FL system.

9.7 LOAD FORECAST DISCUSSION

EPLF is an essential process in the planning of electrical utilities and the operation of the power systems for the utilities. Perfect estimation for load forecasting is the way towards economic saving in operating and maintenance costs. Many studies were done a long time ago and present, where a model of the peak load for the Kingdom of Bahrain to forecast the peak load demand for the upcoming years using polynomial is presented in the present chapter. This type of study helps with increasing the reliability of power supply and delivery system, and making the correct decisions for future development.

Electric peak load demand is the highest recorded demand and is supplied by generators. It is recorded for different time horizons. The estimated and recorded peal load can be on hourly, daily, weekly, monthly, or yearly basis. The EPLF has different characteristics compared to other types of energy commodities. This type of energy should be consumed as soon as it was generated. Electricity Load Forecasting for the peak load that made for various purposes can be classified into three categories. These three categories are short-term, medium-term, and long-term forecasts.

The developing countries like the GCC countries have a peak load growth in electricity mainly based on variables such as economic growth, population, and efficiency standards, coupled with other factors inherent in the mathematical development of forecasting models making accurate projections difficult. The practical outage data statistics collection is helping in the development of methods and models. The study is helping in identifying design variables to reduce the level of risk.

The studies are serving to establish the LTLF, which is used to estimate the load for more than one year, where it is usually used for a period of 20 years, or more years in some cases. The target of this type of forecasting is to plan for the install capacity that helping to build it in the future for the expected load demand based on the projects and the master plan of the country. For example, the LTLF helps the country in allocating and involving the independent power companies to participate in bidding for building power generation or to plan purchasing electrical energy for a long period to manage the energy demand in a most economical and reliable way.

The medium-term load forecast (MTLF) can be used to predict the load from one week to one year. The target of this type of forecasting is to enable electricity utilities and trading companies to estimate the load demand for less than one year and greater than short term load forecast (STLF) period. The MTLF helps the electricity sector companies to negotiate contracts with other companies and to schedule the operations and maintenance.

The STLF is the target to predict the load up to one week. It is used for daily power system operation. The very short-term load forecast (VSTLF) is used for less than one day forecast to meet the load demand during the day in the energy market.

Other important points regarding the estimated electric load study are to avoid the huge in generation capacity in the mid-term and long-term plans, which help the countries to avoid the load shedding and to meet the energy demand in different sector(s). These will help the economic development of the countries.

9.8 LOAD FORECAST ACCURACY TEST

The electric load forecast is applied to different countries. In the present chapter, the electric load study is carried out on the Kingdom of Bahrain, and the curve-fitting using the Polynomial is applied to find suitable formulas for the historical data, and based on this historical data, the estimated load is obtained. Figure 9.1 shows the electric load for the Kingdom of Bahrain through the period of the years from 2003 until 2017, and the formula of the graph is found. In the same way, Figure 9.2 shows the estimated electric load forecast for the Kingdom of Bahrain through the period of the years from 2003 until 2025, and the formula of the graph is obtained.

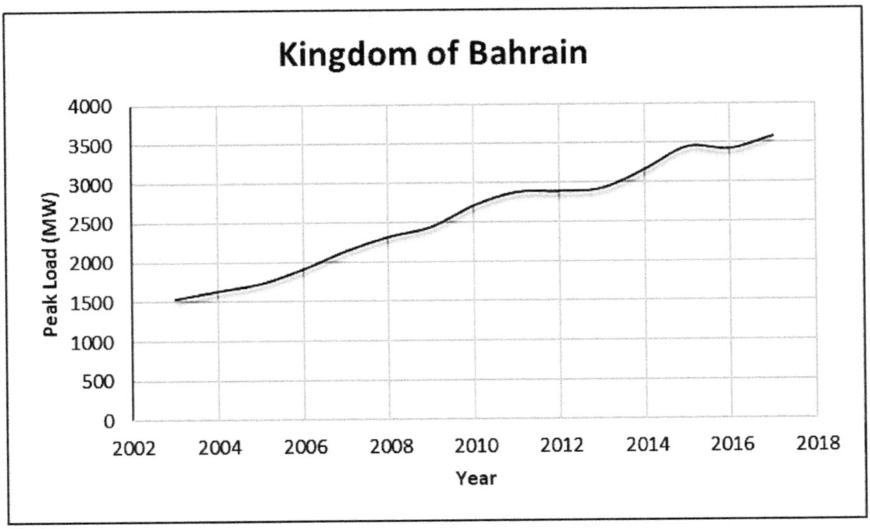

FIGURE 9.1 The actual load for the Kingdom of Bahrain.

The formula for the curve is obtained as:

$$y = 1924.054 + 0.124399x - 0.00007263099x^2 + 1.92115X10\text{–}^8x^3 - 1.846967 \text{ x } 10^{-12}x^4$$

$$a = 1924.054 \pm 57.3$$

$$b = 0.124399 \pm 0.09696$$

$$c = -0.00007263099 \pm 0.0000599$$

$$d = 1.92115 \text{ x } 10^{-8} \pm 1.603 \text{ x } 10^{-8}$$

$$e = -1.846967 \text{ x } 10^{-12} \pm 1.572 \text{ x } 10^{-12}$$

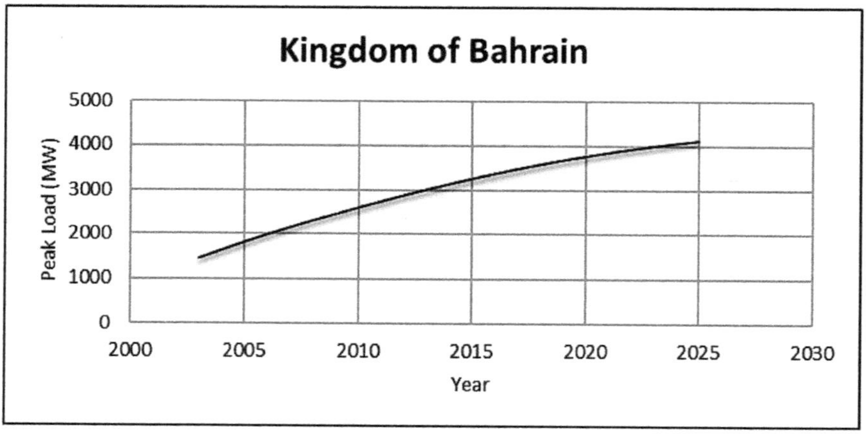

FIGURE 9.2 The estimated load for the Kingdom of Bahrain.

$$y = -12286630 + 12082.81x - 2.96953x^2 - 1.364795 \text{ x } 10^{-10}x^3$$

$$a = -12286630 \pm 2.209$$

$$b = 12082.81 \pm 0.00329$$

$$c = -2.96953 \pm 0.000001633$$

$$d = -1.364795 \text{ x } 10^{-10} \pm 2.703 \text{ x } 10^{-10}$$

9.9 ABSOLUTE PERCENTAGE ERROR

The absolute percent error (APE) measures the size of the error in percentage terms. It is calculated as the average of the unsigned percentage error. The absolute values of the electric load percentage error are calculated and found for the results of polynomial, exponential, and linear techniques—they are illustrated in Chapter 10. Even the shape of the percentage errors is obtained and drawn as presented by the graphs.

KEYWORDS

- **adaptive neuro-fuzzy inference system**
- **artificial neural networks**
- **fuzzy inference system**
- **fuzzy logic**
- **long-term load forecasting**
- **multilayer perceptions**

REFERENCES

Al-Kandari, A. M., Soliman, S. A., & El-Hawary, M. A. Fuzzy short-term electric load forecasting. *Electrical Power and Energy Systems*, **2004**, *26*, 111–122.

He, Y., & Zheng, Y. Short-term power load probability density forecasting based on Yeo-Johnson transformation quantile regression and Gaussian kernel. *Energy*, **2018**, *154*, 143–156.

Hsu, C., & Chen, C. Regional load forecasting in Taiwan applications of artificial neural networks. *Energy Conversion and Management*, **2003**, *44*, 1941–1949.

Islam, S. M., Al-Alawi, S. M., & Ellithy, K. A. Forecasting monthly electric load and energy for a fast growing utility using an artificial neural network. *Electric Power Systems Research*, **1995**, *34*, 1–9.

Jung, H., Song, K., Park, J., & Park, R. Very short-term electric load forecasting for real-time power system operation. *J. Electr. Eng. Technol.*, **2018**, *13*(4), 1419–1424.

Metaxiotis, K., Kagiannas, A., Askounis, D., & Psarras, J. Artificial intelligence in short term electric load forecasting: A state-of-the-art survey for the researcher. *Energy Conversion and Management*, **2003**, *44*, 1525–1534.

Mielczarski, W. *Fuzzy Logic Techniques in Power Systems*, Springer, **1995**, USA.

Oudalov, A., Cherkaoui, R., & Germond, A. J. Application of fuzzy logic techniques for the coordinated power flow control by multiple series FACTS devices. 22nd *IEEE Power Engineering Society International Conference*, **2002**, Sydney, Australia.

Padmakumari, K., Mohandas, K. P., & Thiruvengadam, S. Long term distribution demand forecasting using neuro fuzzy computations. *Electrical Power and Energy Systems*, **1999**, *21*, 315–322.

Papadakis, S. E., Theocharis, J. B., & Bakirtzis, A. G. A load curve based fuzzy modeling technique for short-term load forecasting." *Fuzzy Sets and Systems*, **2003**, *135*, 279–303.
Tamimi, M., & Egbert, R. Short term electric load forecasting via fuzzy neural collaboration. *Electric Power Systems Research*, **2000**, *56*, 243–248.

CHAPTER 10

Power Systems Reliability Applications

10.1 INTRODUCTION

Electrical energy is an essential ingredient for the development of modern society. Almost all aspects of daily life depend on the use of electrical energy and the performance of a power utility. The power utility is measured in terms of the quality and reliability of the supply. Electric power utilities have invested a substantial amount of capital in generating stations, transmission lines, and distribution facilities to supply electrical energy to their customers as economically as possible and with a reasonable assurance of continuity, safety, and quality. This creates the difficult problem of balancing the need for continuity of power supply with the costs involved.

10.2 EQUIVALENT TRANSITION RATE MATRIX

There are a number of methods for constructing the transition rate matrix. For general systems using the Kronecker algebra helps to form a transition rate matrix. This is one method out of a number of methods that help to form a transition rate matrix. Also, the rules and properties can be introduced. The Kronecker algebra is used in this textbook for constructing the transition rate matrix of a system contain a number of power stations.

10.3 TWO-STATE MODEL

The two-state model refers to a generator passing through two-states. In this case, the power generator is passing through Up-State and Down-State. This means the generator is operating and is failing, respectively. The three methods to calculate the transient probabilities for the considered system are discussed earlier in Chapter 8. The results are obtained using the SMM illustrated in Table 10.1 and Figure 10.1. Table 10.2 and Figure 10.2 illustrate the results obtained by the Rung-Kutta method. Finally, Table 10.3 and Figure 10.3 show the results obtained by the Adams method. These three methods are recommended to calculate the transient and steady-state probabilities for any model.

TABLE 10.1 Two-State Model Results Using SMM

Time (year)	P1(t)	P2(t)
0	1	0
0.5	0.9	0.1
1	0.85	0.15
1.5	0.825	0.175
2	0.8125	0.1875
2.5	0.8063	0.1938
3	0.8031	0.1969
3.5	0.8016	0.1984
4	0.8008	0.1992
4.5	0.8004	0.1996
5	0.8002	0.1998
5.5	0.8001	0.1999
6	0.8	0.2
6.5	0.8	0.2
7	0.8	0.2
7.5	0.8	0.2
8	0.8	0.2
15	0.8	0.2
15.5	0.8	0.2
16	0.8	0.2
16.5	0.8	0.2
17	0.8	0.2
17.5	0.8	0.2
18	0.8	0.2
18.5	0.8	0.2
19	0.8	0.2
19.5	0.8	0.2
20	0.8	0.2

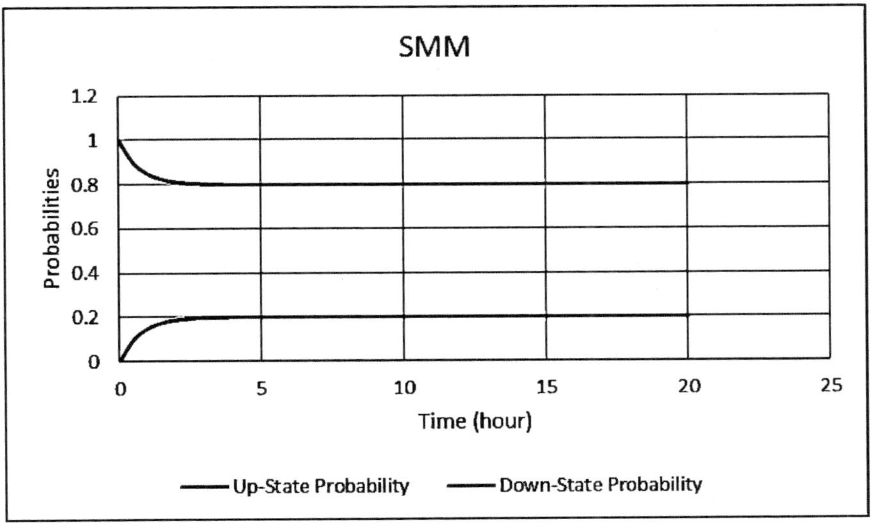

FIGURE 10.1 Two-state model results using SMM.

TABLE 10.2 Two-State Model Results Using Runge-Kutta Method

Time (year)	P1(t)	P2(t)
0	1	0
0.5	0.9245	0.0755
1	0.8775	0.1225
1.5	0.8482	0.1518
2	0.83	0.17
2.5	0.8187	0.1813
3	0.8116	0.1884
3.5	0.8072	0.1928
4	0.8045	0.1955
4.5	0.8028	0.1972
5	0.8017	0.1983
5.5	0.8011	0.1989
6	0.8007	0.1993
6.5	0.8004	0.1996
7	0.8003	0.1997
7.5	0.8002	0.1998
8	0.8001	0.1999
8.5	0.8001	0.1999
9	0.8	0.2
9.5	0.8	0.2

TABLE 10.2 *(Continued)*

Time (year)	P1(t)	P2(t)
10	0.8	0.2
10.5	0.8	0.2
11	0.8	0.2
11.5	0.8	0.2
12	0.8	0.2
12.5	0.8	0.2
13	0.8	0.2
13.5	0.8	0.2
14	0.8	0.2
14.5	0.8	0.2
15	0.8	0.2
15.5	0.8	0.2
16	0.8	0.2
16.5	0.8	0.2
17	0.8	0.2
17.5	0.8	0.2
18	0.8	0.2
18.5	0.8	0.2
19	0.8	0.2
19.5	0.8	0.2
20	0.8	0.2

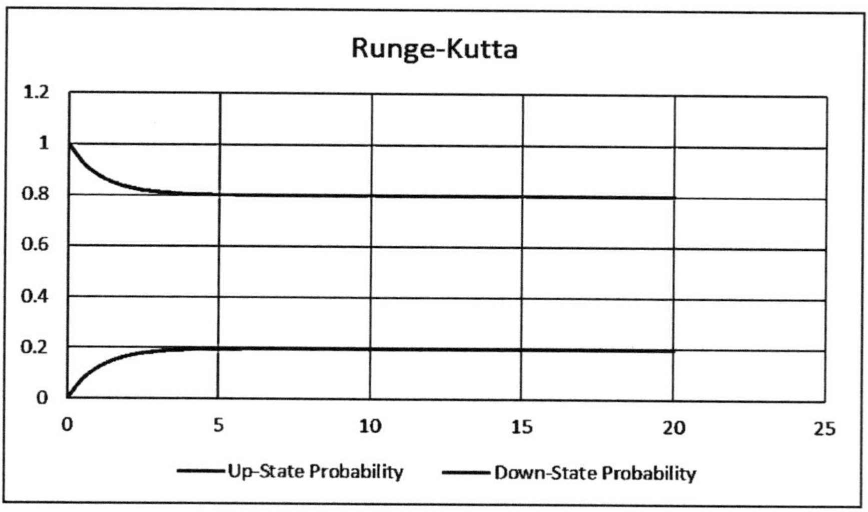

FIGURE 10.2 Two-state model results using the Runge-Kutta method.

TABLE 10.3 Two-State Model Results Using Adams Method

Time (hr)	P1(t)	P2(t)
0	1	0
0.5	0.91	0.09
1	0.86	0.14
1.5	0.83	0.17
2	0.815	0.185
2.5	0.805	0.195
3	0.8	0.2
3.5	0.8	0.2
4	0.8	0.2
4.5	0.8	0.2
5	0.8	0.2
5.5	0.8	0.2
6	0.8	0.2
6.5	0.8	0.2
7	0.8	0.2
7.5	0.8	0.2
8	0.8	0.2
8.5	0.8	0.2
9	0.8	0.2
9.5	0.8	0.2
10	0.8	0.2
11	0.8	0.2
11.5	0.8	0.2
12	0.8	0.2
13	0.8	0.2
14	0.8	0.2
15	0.8	0.2
16	0.8	0.2
17	0.8	0.2
18	0.8	0.2
19	0.8	0.2
20	0.8	0.2

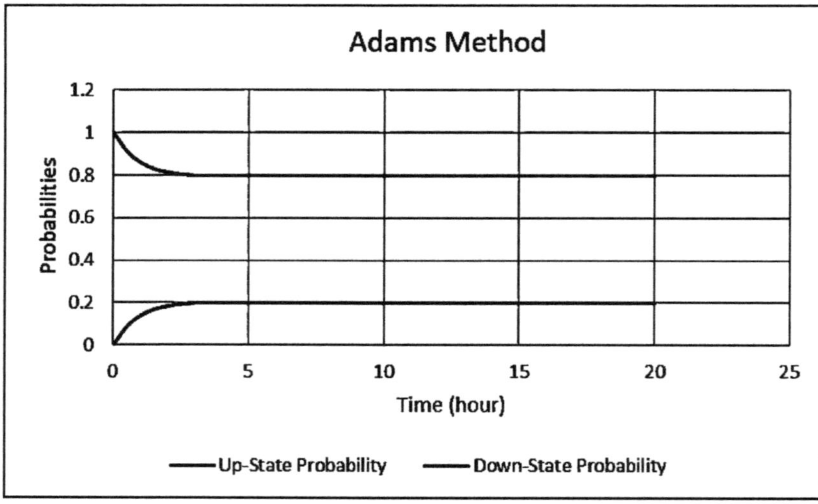

FIGURE 10.3 Two-state model results using the Adams method.

The curve fitting technique, applied to the three methods, is used to find out the results for the two-state model. The results are illustrated in Figures 10.4a, b using SMM, Figures 10.5a, b using the Runge-Kutta method, and Figures 10.6a, b using Adams method. The suitable equations for the three methods are summarized in Table 10.4.

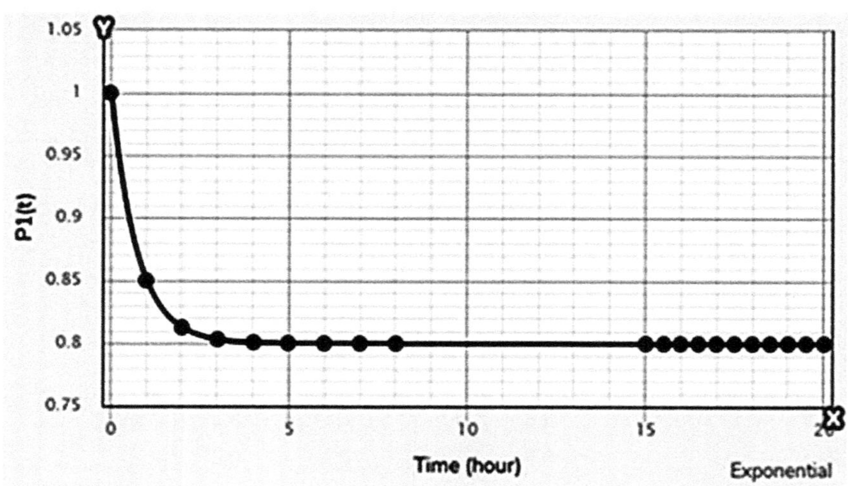

FIGURE 10.4a First probability using SMM.

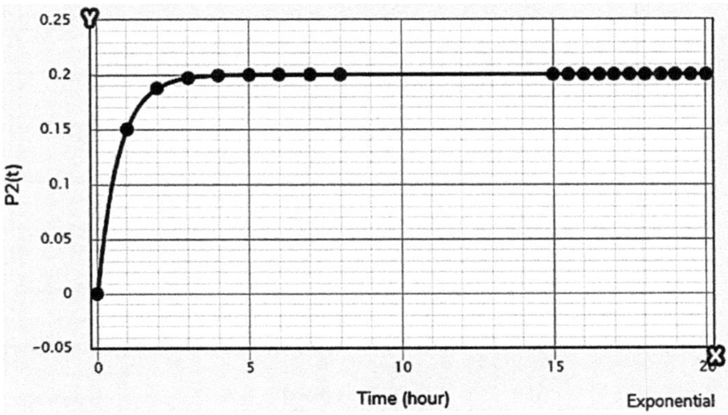

FIGURE 10.4b Second probability using SMM.

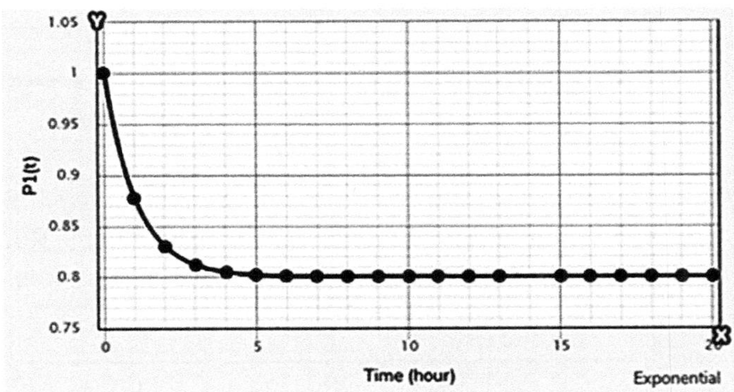

FIGURE 10.5a First probability using the Runge-Kutta method.

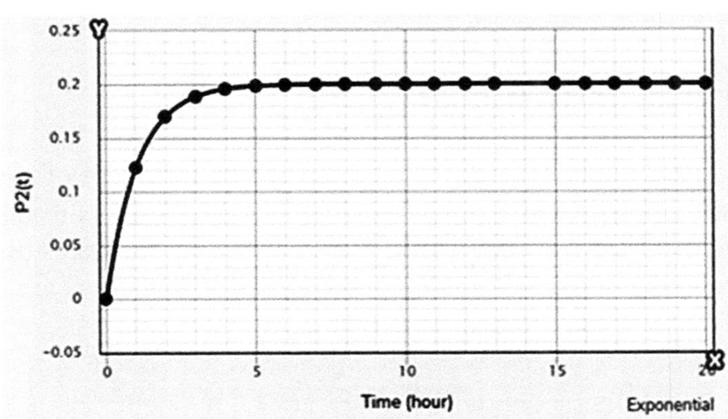

FIGURE 10.5b Second probability using the Runge-Kutta method.

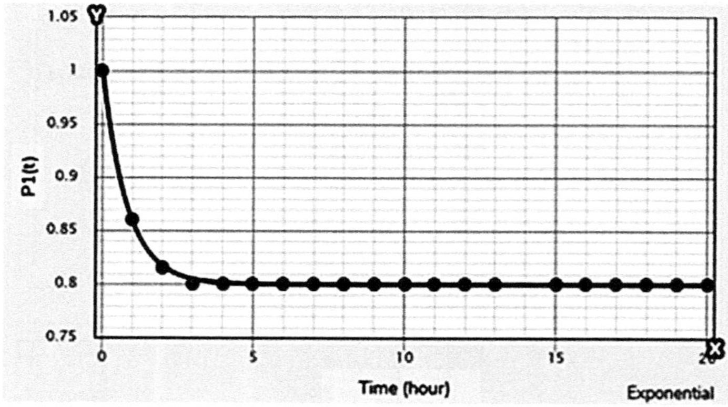

FIGURE 10.6a First probability using the Adams method.

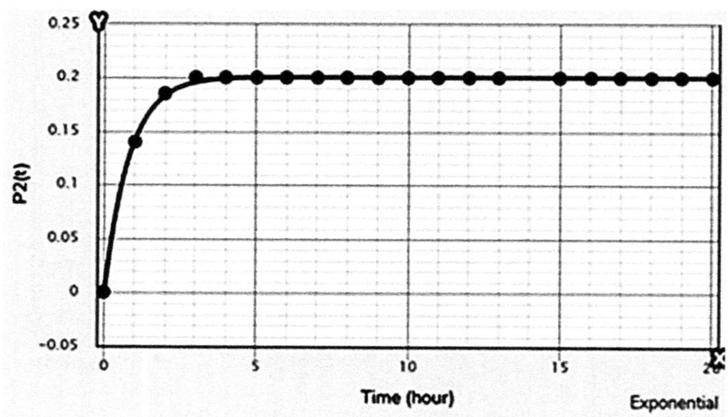

FIGURE 10.6b Second probability using the Adams method.

10.4 THREE-STATE MODELS

The three-state model (Figure 10.7) is considered for a case of two generators under operation in a power station. The first state considered that both generators are working, where the second state illustrates that one is working and the others fail. In the third state, both generators are in a fail-state. Using the Laplace transforms will result:

$$P_1(t)=\frac{\lambda^2}{(\lambda+\mu)^2}e^{-2(\mu+\lambda)t}+\frac{2\lambda\mu}{(\lambda+\mu)^2}e^{-(\mu+\lambda)t}+\frac{\mu^2}{(\lambda+\mu)^2}$$

TABLE 10.4 Two-State Results Curve Fitting, $y = a + b\,e^{-cx}$

Method	Probability	a	b	c
SMM	1	0.7999963±0.000003416	0.2000039±0.00001428	1.386269±0.0002655
	2	0.2000037±0.000003416	–0.2000039±0.00001428	1.386269±0.0002655
Runge-Kutta	1	0.7999955±0.000005142	0.2000096±0.00002047	0.948314±0.000219
	2	0.2000045±0.000005142	–0.2000096±0.00002047	0.948314±0.000219
Adams	1	0.7996066±0.0003309	0.2007632±0.001368	1.236986±0.02122
	2	0.2003934±0.0003309	–0.2007631±0.001368	1.236985±0.02122

$$P_2(t) = \frac{2\mu\lambda}{(\lambda+\mu)^2} + \frac{2\lambda(\lambda-\mu)}{(\lambda+\mu)^2} e^{-(\mu+\lambda)t} - 2\frac{\lambda^2}{(\lambda+\mu)^2} e^{-2(\mu+\lambda)t}$$

$$P_3(t) = \frac{\lambda^2}{(\lambda+\mu)^2} e^{-2(\mu+\lambda)t} - \frac{2\lambda^2}{(\lambda+\mu)^2} e^{-(\mu+\lambda)t} + \frac{\lambda^2}{(\lambda+\mu)^2}$$

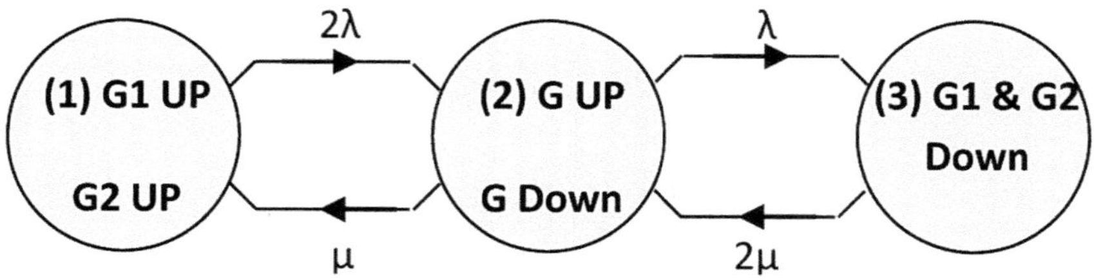

FIGURE 10.7 Three-state of the two-generation model.
Source: Modified from Billinton and Allan, 1992.

The results are illustrated in three cases. These cases are summarized in Tables 10.5–10.8 and Figures 10.8–10.10.

TABLE 10.5 Results Summary for Three Cases Using Laplace Transforms

Case #	μ (per year)	λ (per year)	Table	Figure
1	0.9	0.1	10.5	10.5
2	0.85	0.15	10.6	10.6
3	0.75	0.25	10.7	10.7

TABLE 10.6 3-State Model Using Laplace Transforms, μ = 0.9 per year, λ = 0.1 per year

TIME (YEAR)	P1(T)	P2(T)	P3(T)
0	1	0	0
1	0.877571652	0.118432584	0.003995764
2	0.834543507	0.157980042	0.007476451
3	0.81898646	0.171984494	0.009029046
4	0.81330017	0.177062789	0.009637042
5	0.811213284	0.17892102	0.009865695
6	0.810446237	0.179603277	0.009950486
7	0.810164147	0.179854082	0.009981771
8	0.810060384	0.179946324	0.009993292
9	0.810022214	0.179980254	0.009997532

TABLE 10.6 *(Continued)*

TIME (YEAR)	P1(T)	P2(T)	P3(T)
10	0.810008172	0.179992736	0.009999092
11	0.810003006	0.179997328	0.009999666
12	0.810001106	0.179999017	0.009999877
13	0.810000407	0.179999638	0.009999955
14	0.81000015	0.179999867	0.009999983
15	0.810000055	0.179999951	0.009999994
16	0.81000002	0.179999982	0.009999998
17	0.810000007	0.179999993	0.009999999
18	0.810000003	0.179999998	0.01
19	0.810000001	0.179999999	0.01
20	0.81	0.18	0.01
21	0.81	0.18	0.01
22	0.81	0.18	0.01
23	0.81	0.18	0.01
24	0.81	0.18	0.01
25	0.81	0.18	0.01
26	0.81	0.18	0.01
27	0.81	0.18	0.01
28	0.81	0.18	0.01
29	0.81	0.18	0.01
30	0.81	0.18	0.01

TABLE 10.7 Three-State Model, $\mu = 0.85$ per year, $\lambda = 0.15$ per year

Time (year)	P1(t)	P2(t)	P3(t)
0	1	0	0
1	0.819354301	0.17165523	0.008990469
2	0.757422599	0.225755387	0.016822014
3	0.735251474	0.244433172	0.020315354
4	0.727178036	0.25113862	0.021683344
5	0.724219198	0.253582988	0.022197814
6	0.72313222	0.254479186	0.022388594
7	0.722732549	0.254808467	0.022458984
8	0.722585546	0.254929548	0.022484907
9	0.72253147	0.254974083	0.022494447
10	0.722511577	0.254990466	0.022497957
11	0.722504259	0.254996493	0.022499248
12	0.722501567	0.25499871	0.022499724
13	0.722500576	0.254999525	0.022499898
14	0.722500212	0.254999825	0.022499963

TABLE 10.7 *(Continued)*

Time (year)	P1(t)	P2(t)	P3(t)
15	0.722500078	0.254999936	0.022499986
16	0.722500029	0.254999976	0.022499995
17	0.722500011	0.254999991	0.022499998
18	0.722500004	0.254999997	0.022499999
19	0.722500001	0.254999999	0.0225
20	0.722500001	0.255	0.0225
21	0.7225	0.255	0.0225
22	0.7225	0.255	0.0225
23	0.7225	0.255	0.0225
24	0.7225	0.255	0.0225
25	0.7225	0.255	0.0225
26	0.7225	0.255	0.0225
27	0.7225	0.255	0.0225
28	0.7225	0.255	0.0225
29	0.7225	0.255	0.0225
30	0.7225	0.255	0.0225

TABLE 10.8 Three-State Model, $\mu = 0.75$ per year, $\lambda = 0.25$ per year

Time (year)	P1(t)	P2(t)	P3(t)
0	1	0	0
1	0.708913246	0.266113229	0.024973525
2	0.614395459	0.338876724	0.046727817
3	0.581325073	0.362243389	0.056431538
4	0.569389331	0.370379157	0.060231512
5	0.565029568	0.373309838	0.061660594
6	0.563429916	0.374379544	0.06219054
7	0.562842008	0.374771926	0.062386067
8	0.562625806	0.37491612	0.062458074
9	0.56254628	0.374969146	0.062484575
10	0.562517025	0.37498865	0.062494325
11	0.562506263	0.374995825	0.062497912
12	0.562502304	0.374998464	0.062499232
13	0.562500848	0.374999435	0.062499717
14	0.562500312	0.374999792	0.062499896
15	0.562500115	0.374999924	0.062499962
16	0.562500042	0.374999972	0.062499986
17	0.562500016	0.37499999	0.062499995
18	0.562500006	0.374999996	0.062499998
19	0.562500002	0.374999999	0.062499999

TABLE 10.8 *(Continued)*

Time (year)	P1(t)	P2(t)	P3(t)
20	0.562500001	0.374999999	0.0625
21	0.5625	0.375	0.0625
22	0.5625	0.375	0.0625
23	0.5625	0.375	0.0625
24	0.5625	0.375	0.0625
25	0.5625	0.375	0.0625
26	0.5625	0.375	0.0625
27	0.5625	0.375	0.0625
28	0.5625	0.375	0.0625
29	0.5625	0.375	0.0625
30	0.5625	0.375	0.0625

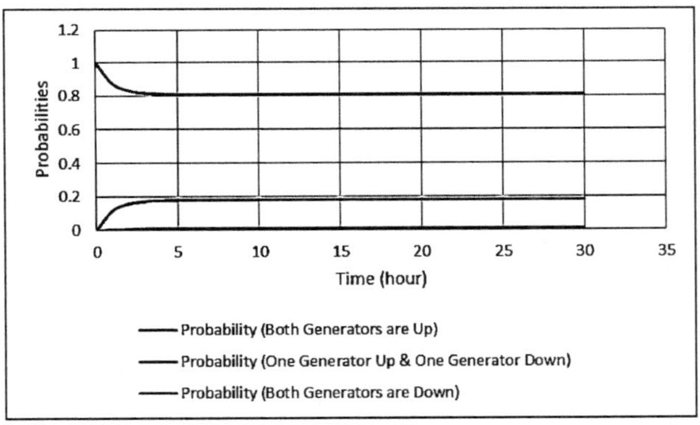

FIGURE 10.8 Three-state model, μ = 0.9 per year, λ = 0.1 per year.

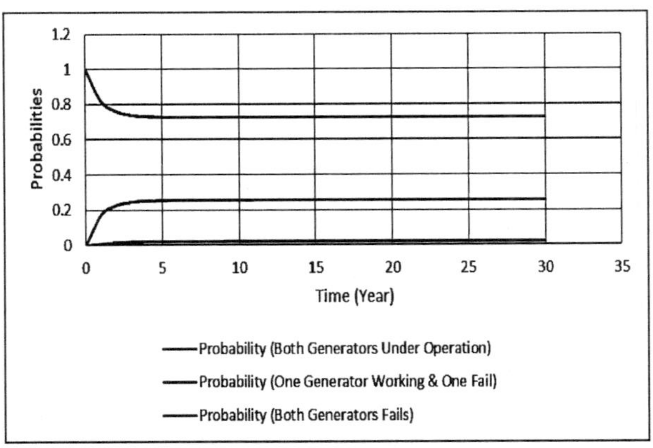

FIGURE 10.9 Three-state model, μ = 0.85 per year, λ = 0.15 per year.

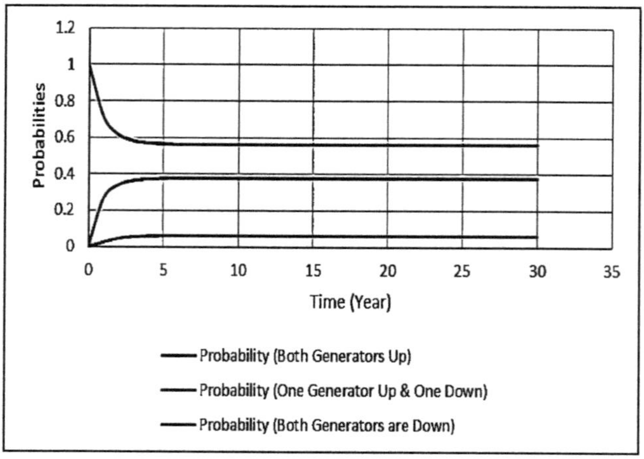

FIGURE 10.10 Three-state model, $\mu = 0.75$ per year, $\lambda = 0.25$ per year.

The Laplace Transforms and Adams Methods have been applied to obtain the transient probabilities shown in the present chapter. The curve fitting technique applied to the three cases of the three-state model. The results obtained are summarized in Figures 10.11a,b,c, 10.12a,b,c, and 10.13a,b,c as the values of the repair and failure rates printed. The results of the fitting equations for the three cases are summarized in Table 10.9.

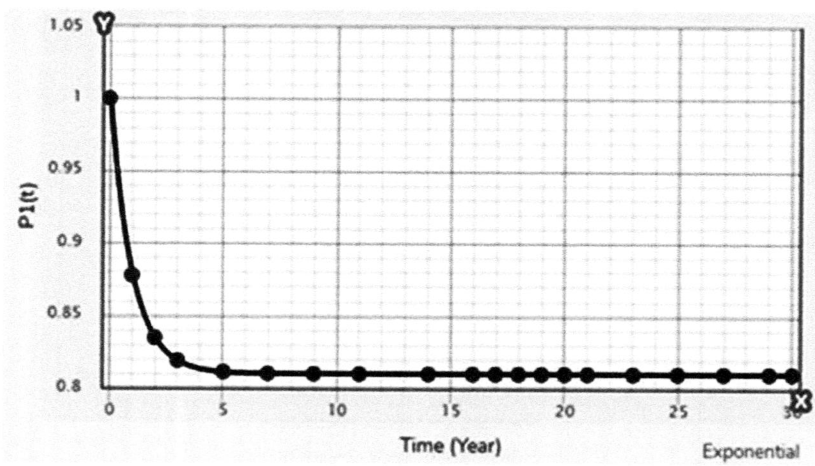

FIGURE 10.11a The first probability for the three-state model using Laplace transforms ($\mu = 0.9$, $\lambda = 0.1$).

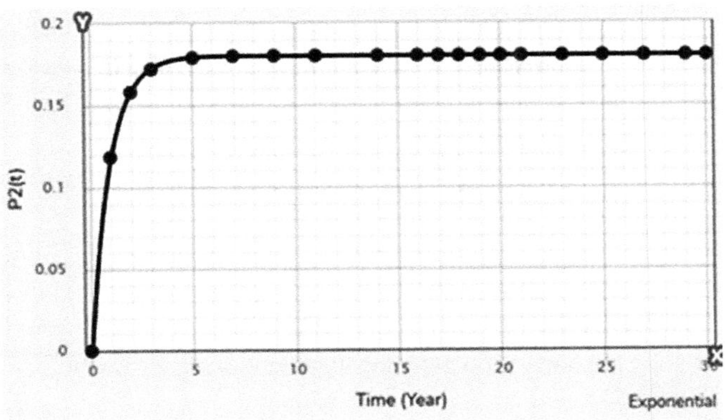

FIGURE 10.11b The second probability for the three-state model using Laplace transforms ($\mu = 0.9, \lambda = 0.1$).

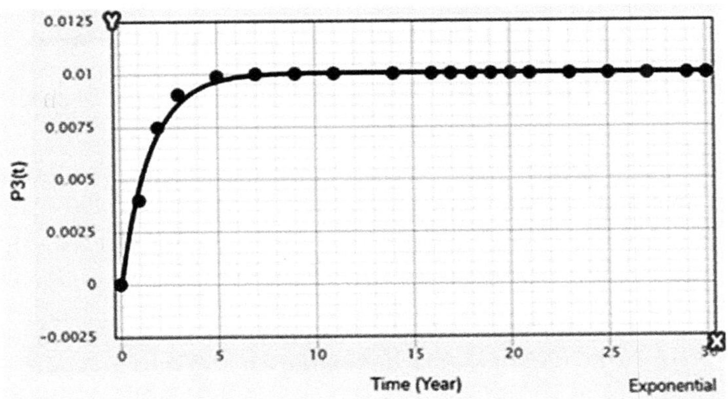

FIGURE 10.11c The third probability for the three-state model using Laplace transforms ($\mu = 0.9, \lambda = 0.1$).

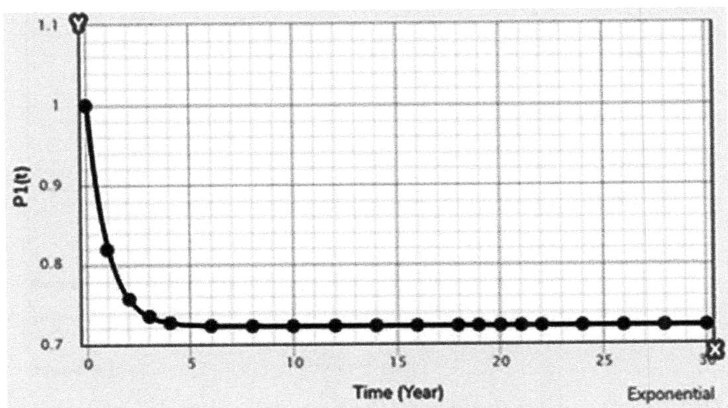

FIGURE 10.12a The first probability for the three-state model using Laplace transforms ($\mu = 0.85, \lambda = 0.15$).

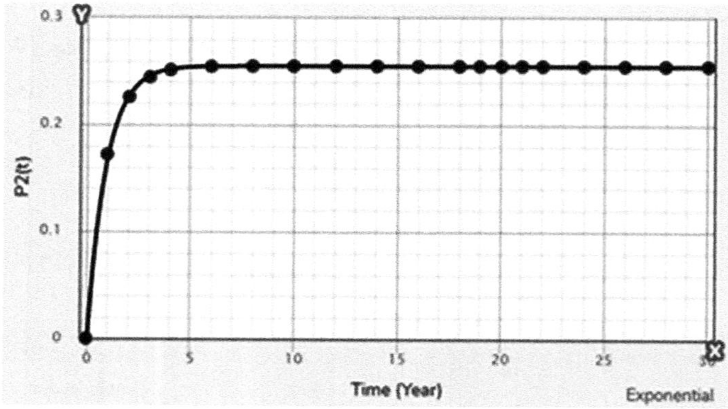

FIGURE 10.12b The second probability for the three-state model using Laplace transforms ($\mu = 0.85$, $\lambda = 0.15$).

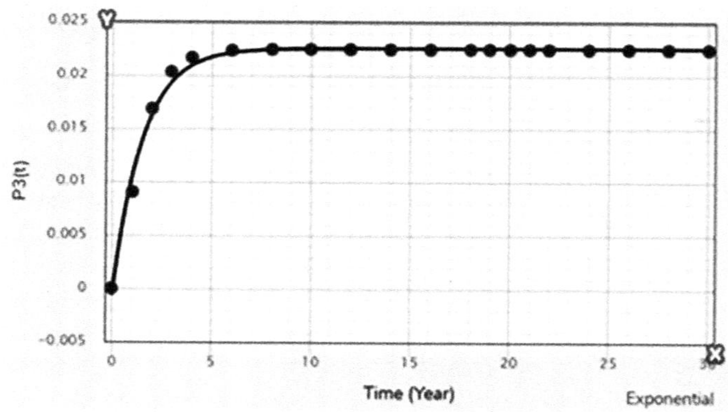

FIGURE 10.12c The third probability for the three-state model using Laplace transforms ($\mu = 0.85$, $\lambda = 0.15$).

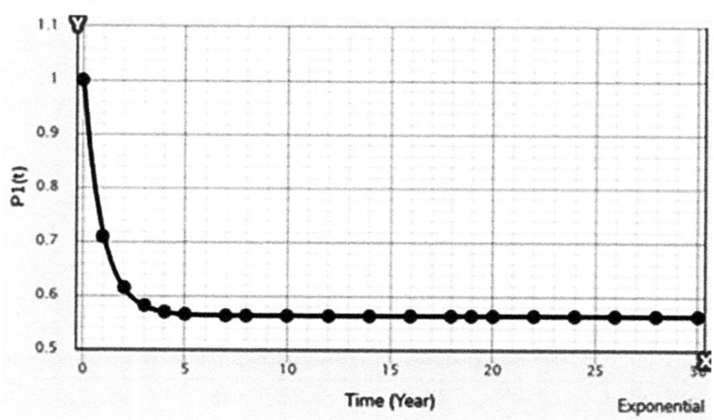

FIGURE 10.13a The first probability for the three-state model using Laplace transforms ($\mu = 0.75$, $\lambda = 0.25$).

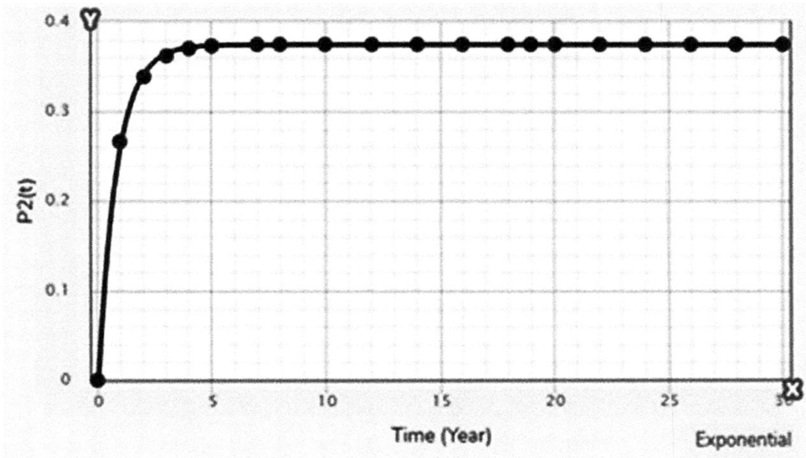

FIGURE 10.13b The second probability for the three-state model using Laplace transforms ($\mu = 0.75$, $\lambda = 0.25$).

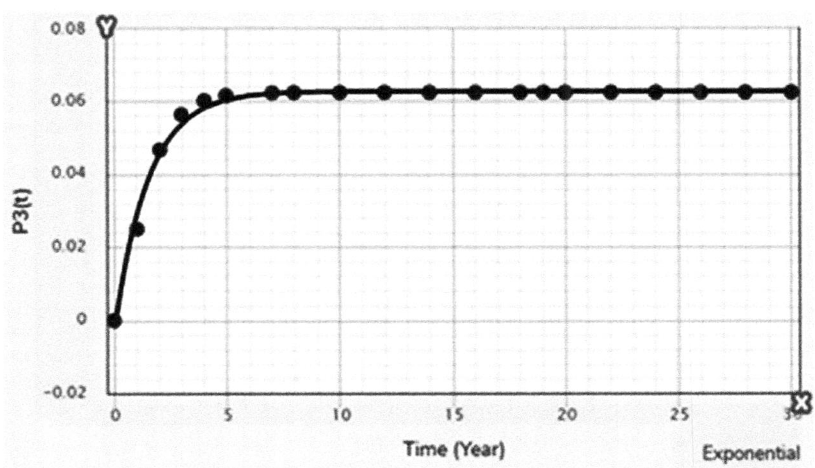

FIGURE 10.13c The third probability for the three-state model using Laplace transforms ($\mu = 0.75$, $\lambda = 0.25$).

10.5 SYSTEM COMBINATION

Any number of Markov models (representing a number of systems) can form one large system representing an electric power station. This clear when the Kronecker multiplication is used.

TABLE 10.9 Three-State Results Curve Fitting Using Laplace Transforms, $y = a + b\,e^{-cx}$

Method	Probability	a	b	c
LT,	1	0.8100282±0.00003175	0.1899086±0.0001303	1.029159±0.001632
$\mu = 0.9$	2	0.1799455±0.00005997	−0.1798336±0.0002469	1.063473±0.003417
$\lambda = 0.1$	3	0.01003722±0.00005733	−0.01026308±0.00005997	0.6513805±0.02942
LT,	1	0.7225763±0.00007297	0.277281±0.0002973	1.045382±0.002606
$\mu = 0.85$	2	0.254857±0.0001317	−0.2546274±0.00054	1.104067±0.00557
$\lambda = 0.15$	3	0.0225983±0.0001371	−0.02314623±0.0005211	0.6590729±0.03037
LT,	1	0.5627287±0.0001916	0.4369243±0.0007806	1.082793±0.004574
$\mu = 0.75$	2	0.3746189±0.0003001	−0.3742009±0.001239	1.215129±0.01003
$\lambda = 0.25$	3	0.06282769±0.0003926	−0.06439482±0.001466	0.6609917±0.03079

10.6 PROPOSED MODEL ADVANTAGE

The Kronecker product is very efficient to form a large model of the electric power system, where this an advantage recorded for the Kronecker product.

10.7 MEMBERSHIP FUNCTION

Figure 10.14 illustrates the four layers connection for three inputs and one output. These four layers represent the Neuro-Fuzzy model, which is used to represent the developed estimated peak load model in this study. It is clear that the results obtained automatically through the training of data performed in the Adaptive Neuro-Fuzzy Inference System (ANFIS).

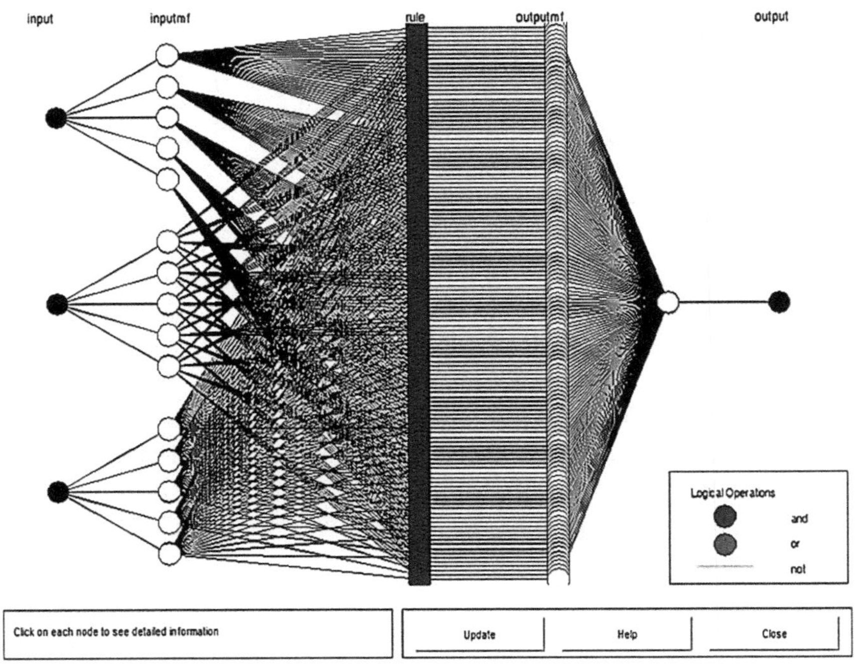

FIGURE 10.14 (See color insert.) ANFIS structure with three inputs and one output.
Source: Reprinted from Qamber and Al-Hamad, 2016.

Figure 10.15 shows the proposed model using the Neuro-Fuzzy, where the proposed model is three inputs and one output model. The three inputs are the year, population, and GDP. The output is the peak load for the country. Therefore, the three inputs used as follows:

1. **Input 1:** Year.
2. **Input 2:** Population.
3. **Input 3:** GDP.

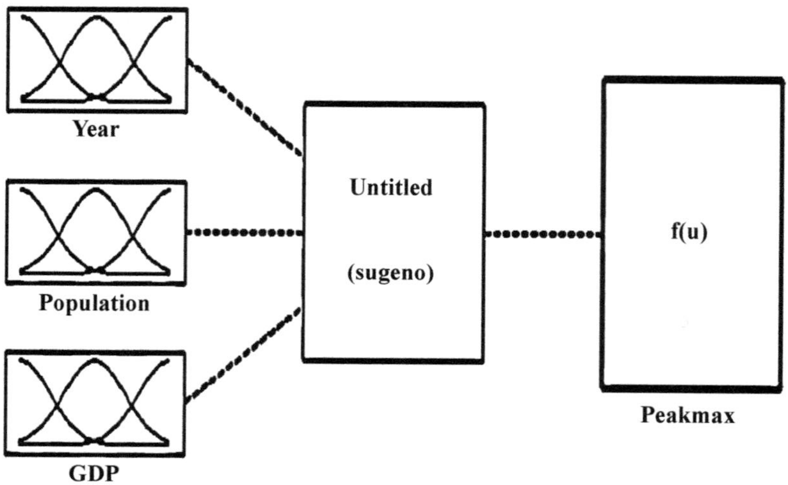

FIGURE 10.15 (See color insert.) Artificial neuro-fuzzy logic model using long term load forecasting.
Source: Reprinted from Qamber and Al-Hamad, 2016.

Figure 10.16 illustrates the actual data collected and presented for member state's peak demand, where the estimated peak load demand for the coming years generated by Neuro-Fuzzy model. Figure 10.17 membership function used in the neuro-fuzzy logic model for long term load forecasting.

10.7.1 VALIDATION OF MEMBERSHIP FUNCTIONS

The ANFIS or adaptive network-based fuzzy inference system can be a shortage as "ANFIS." ANFIS is a kind of artificial neural network-modeling standard. It was adopted to effectively tune the membership function to decrease the output error and maximize performance indicators. At the same time, ANFIS Editor Display made up four types. These types are Load data, General is, Train FIS, and Test FIS. The load data is used for training, testing, and checking. In the research, the ANFIS approach is recommended, where it is used by adding the Artificial Neural Network to the Fuzzy System, which has successfully simulated for the load forecasting.

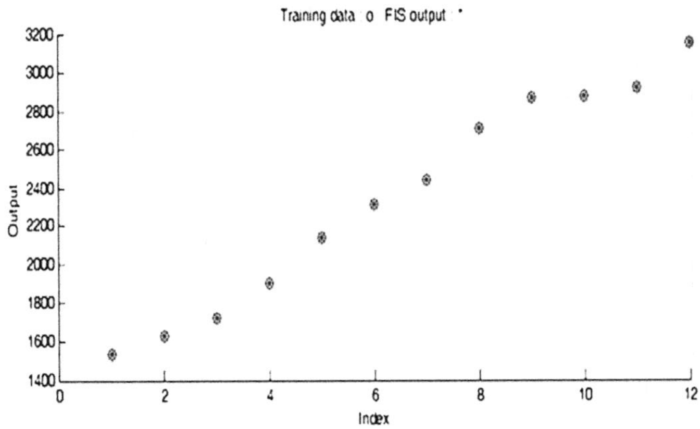

FIGURE 10.16 (See color insert.) Actual data used versus the output of the neuro-fuzzy model.

Source: Reprinted from Qamber and Al-Hamad, 2016.

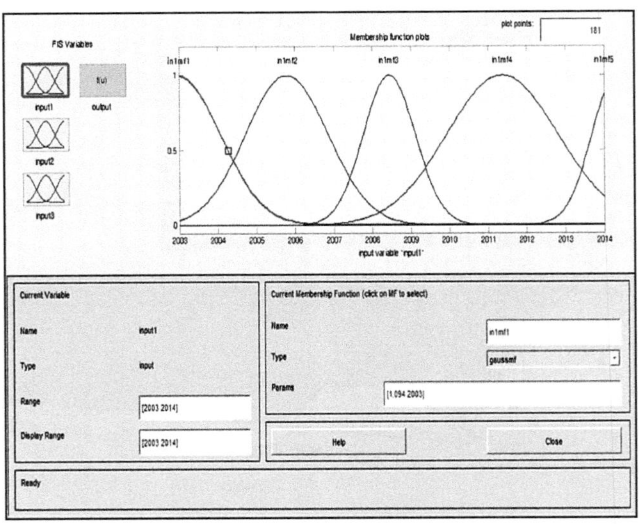

FIGURE 10.17 (See color insert.) Membership function used in the neuro-fuzzy logic model for long term load forecasting.

Source: Reprinted from Qamber and Al-Hamad, 2016.

10.7.2 LOAD FORECASTED SURFACE

The electricity load forecasting is an active research topic with important practical effects for almost any industry. It is well known that the load forecasting is a technique used by electric power or electric energy-providing companies to estimate the power or energy needed to satisfy the demand and supply stability. The accuracy of forecasting is of great worth for the operational and managerial loading of a utility company. The most accurate estimation of energy consumption and supplies has a helpful impact on active budgets. In the instance of the electrical industry, accurate electrical load estimating is a suitable means to provide well-grounded information and capable energy management for arrangement in both the mid and long term and for the network operation in the short term. Back to the term (surface). The term (surface) is viewer represented by a graphical interface that lets the researcher examines the output surface of the fuzzy inference system (FIS) for two inputs. Figure 10.18 illustrates an example of a load forecasting of a country. The inputs are the population and years to find the estimates load for further years, as shown in Figure 10.18.

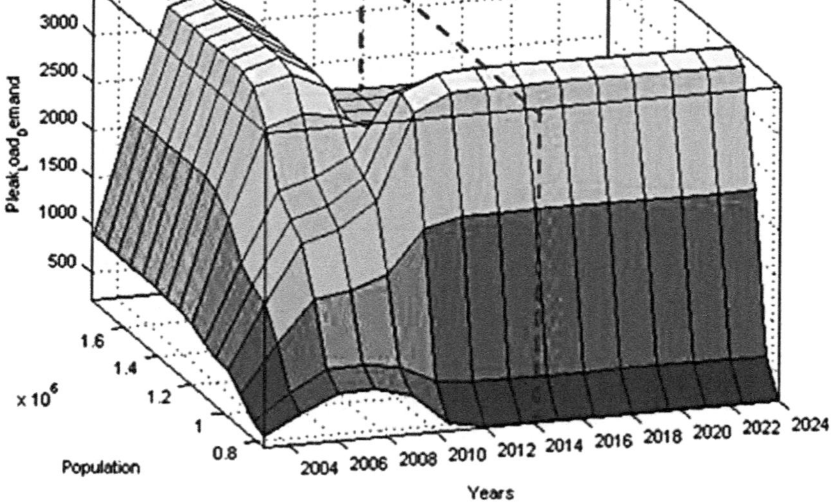

FIGURE 10.18 (See color insert.) A surface graph for a country, estimating the load.
Source: Reprinted from Qamber and Al-Hamad, 2016.

10.8 THE EFFECT OF TEMPERATURE

To design high-precision estimators for a load-cell-based weighting system with temperature compensation, ANFIS is used to model the relationship with actual weights of samples. In addition, ANFIS can improve the precision of load cell. ANFIS is used to model the relationship between the reading of load cells and the actual weight of samples considering the temperature-varying effect and nonlinearity of the load cells. The model of the load-cell-based weighting system can accurately estimate the weight of test samples from the load cell reading. The proposed ANFIS-based method is convenient to use and can improve the 117 precision of digital load cell measurement systems. The temperature and humidity are two parameters considered. The humidity and the temperature data are fed to the fuzzy logic (FL) system, where the output is the load. The load is proportional to these two parameters. The neural networks used for training and comparing the set of the past load to predict the future. In this case, the load does not just depend on the temperature and humidity but also depends on the other data like a number of customers for Load Forecasting of the Power System Planning using Fuzzy-Neural.

10.8.1 FIRST MODEL

There are a number of methods used to estimate the electric load for the countries. Kingdom of Bahrain taken as an example in the present chapter to estimate its electric load for the future up to the year 2025 using the polynomial. Table 10.10 shows the estimated peak loads for the Kingdom of Bahrain with calculated results using the Polynomial. The electric load percentage error using the polynomial is calculated and found as 2.471288523. The estimated load versus the years is plotted and illustrated in Figure 10.19. The percentage error of the peak load for the results of the polynomial is shown in Figure 10.20. The results of the percentage error for the peak load converted to absolute values and shown in Figure 10.21. Furthermore, when the square of the peak load error versus the years obtained, which is shown in Figure 10.22, the shape of that becoming almost a sinusoidal.

10.8.2 SECOND MODEL

There are a number of methods used to estimate the electric load for the countries. Kingdom of Bahrain taken as an example in the present chapter to estimate its electric load for the future up to the year 2025 using the exponential. Table 10.11 shows the peak loads for the Kingdom of Bahrain with calculated estimated peak loads results using the Exponential. The electric load percentage error using the exponential is calculated and found as5.100549189. The estimated load versus the years is plotted and illustrated in Figure 10.23. The percentage error of the peak load for the results of the exponential is shown in Figure 10.24. The results of the absolute value percentage error for the peak load is found and shown in Figure 10.25. Furthermore, when the square of the peak load error versus the years obtained, which is shown in Figure 10.26, the shape of that becoming almost a sinusoidal.

TABLE 10.10 The Estimated Peak Loads for the Kingdom of Bahrain with Calculated Results Using the Polynomial

Year	Peak Load Actual (MW)	Peak Load Calculated (MW)	% Peak Load Error
2003	1535	1449.329226	5.581157948
2004	1632	1633.227512	–0.075215196
2005	1725	1811.186738	–4.996332609
2006	1906	1983.206902	–4.050729381
2007	2136	2149.288006	–0.622097636
2008	2314	2309.430048	0.197491443
2009	2438	2463.63303	–1.051395796
2010	2708	2611.89695	3.54885709
2011	2871	2754.22181	4.067509248
2012	2880	2890.607608	–0.368319722
2013	2917	3021.054346	–3.567169883
2014	3152	3145.562022	0.204250571
2015	3441	3264.130638	5.140057033
2016	3415	3376.760192	1.119760117
2017	3572	3483.450686	2.478984169
2018		3584.202118	
2019		3679.01449	
2020		3767.8878	
2021		3850.82205	
2022		3927.817238	
2023		3998.873366	
2024		4063.990432	
2025		4123.168438	

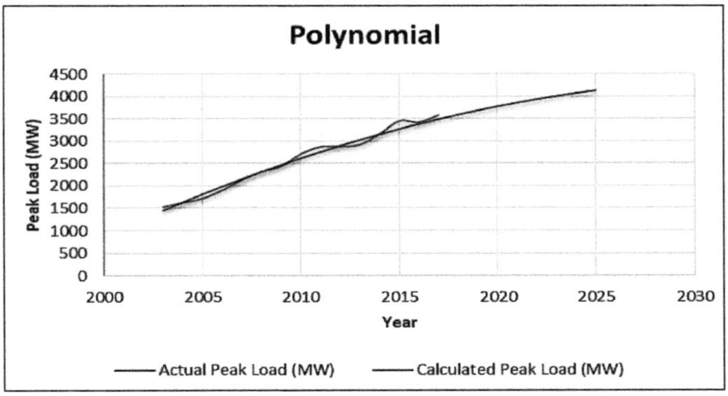

FIGURE 10.19 The estimated load versus the years.

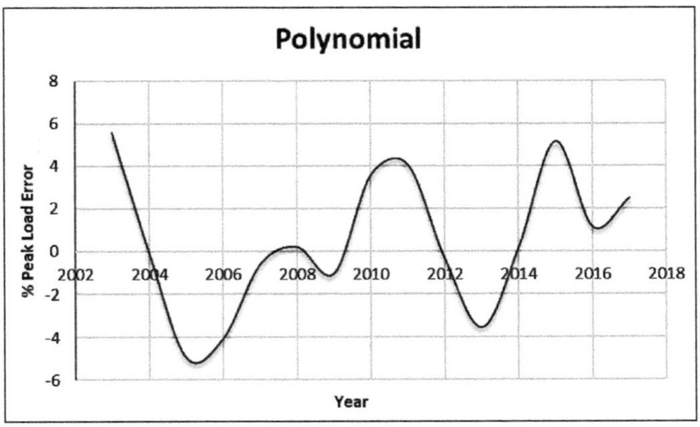

FIGURE 10.20 The percentage error for the load versus the years.

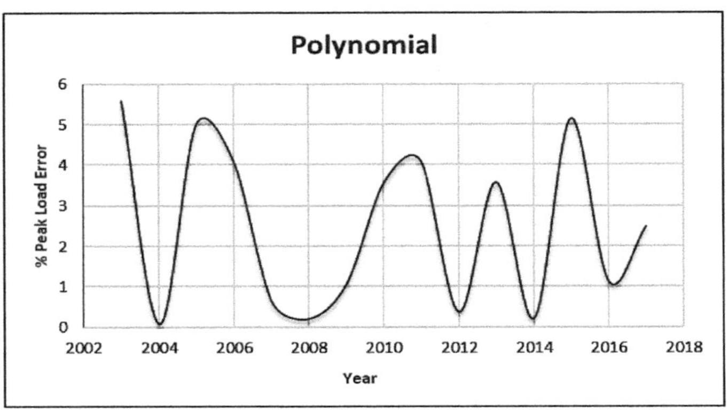

FIGURE 10.21 The absolute value of percentage error for the load versus the years.

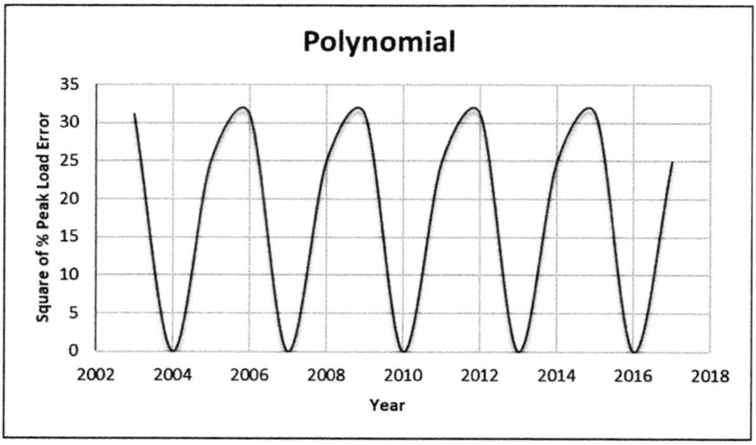

FIGURE 10.22 The square of the percentage error for the load versus the years.

TABLE 10.11 The Estimated Peak Loads for the Kingdom of Bahrain with Calculated Results Using the Exponential

Year	Actual Peak Load (MW)	Estimated Peak Load (MW)	% Error
2003	1535	1570.298109	–2.299551106
2004	1632	1681.342525	–3.023439059
2005	1725	1800.239503	–4.361710307
2006	1906	1927.544339	–1.130343093
2007	2136	2063.851601	3.377734032
2008	2314	2209.797899	4.503115861
2009	2438	2366.064862	2.950579889
2010	2708	2533.382322	6.448215581
2011	2871	2712.53172	5.519619656
2012	2880	2904.349756	–0.845477656
2013	2917	3109.732302	–6.60720952
2014	3152	3329.638576	–5.635741633
2015	3441	3565.095633	–3.606382826
2016	3415	3817.203153	–11.77754475
2017	3572	4087.138583	–14.42157287
2018		4376.162631	
2019		4685.62516	
2020		5016.971486	
2021		5371.749133	
2022		5751.615059	
2023		6158.343393	
2024		6593.833726	
2025		7060.119975	

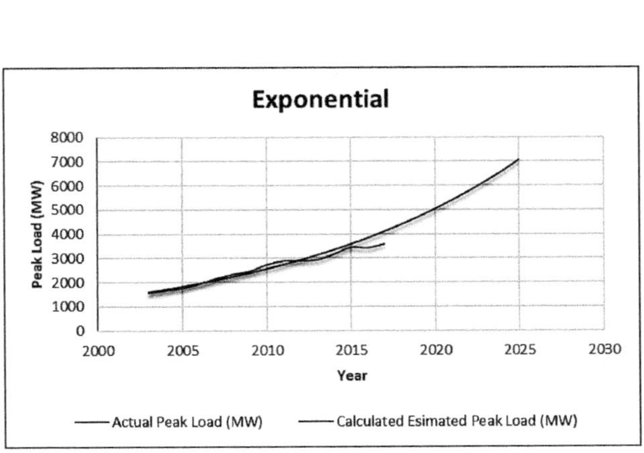

FIGURE 10.23 The estimated load versus the years.

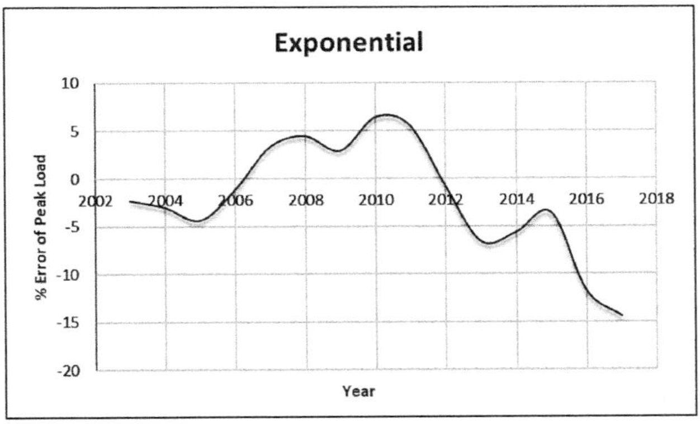

FIGURE 10.24 The percentage error for the load versus the years.

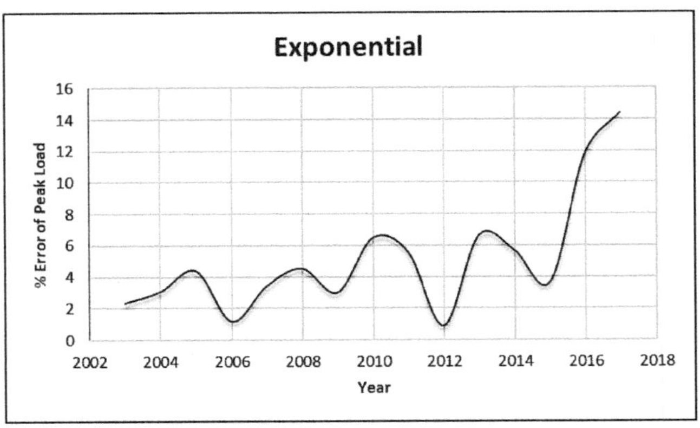

FIGURE 10.25 The absolute value of percentage error for the load versus the years.

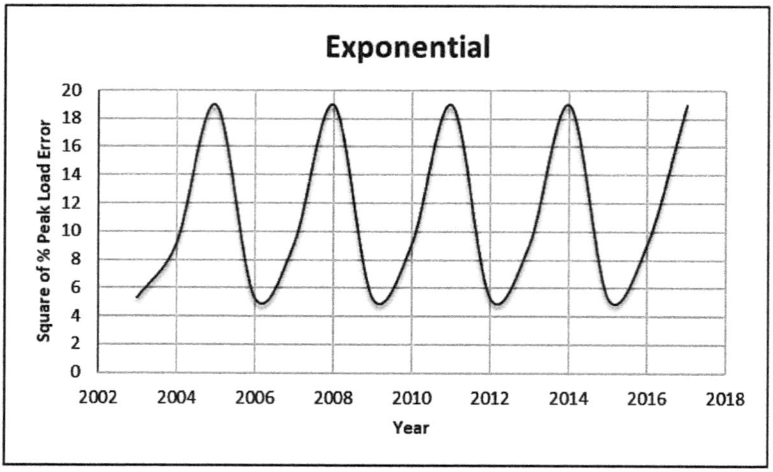

FIGURE 10.26 The square of the percentage error for the load versus the years.

10.8.3 THIRD MODEL

There are a number of methods used to estimate the electric load for the countries. Kingdom of Bahrain taken as an example in the present chapter to estimate its electric load for the future up to the year 2025 using the linear. Table 10.12 shows the peak loads for the Kingdom of Bahrain with calculated results using the Linear. The electric load percentage error using the linear is calculated and found as 2.533884781. The estimated load versus the years is plotted and illustrated in Figure 10.27. The percentage error of the peak load for the results of the linear is shown in Figure 10.28. The results of the percentage error for the absolute value peak load are found and shown in Figure 10.29. Furthermore, when the square of the peak load error versus the years obtained, which is shown in Figure 10.30, the shape of that becoming almost a semi-sinusoidal.

10.9 COMPARISON BETWEEN THE MODELS

Three techniques were used to find the estimated peak load for the Kingdom of Bahrain using the polynomial, exponential, and linear. The average percentage error was determined and found to be 2.471288523 using the polynomial technique. The average percentage errors were determined and found to be 5.100549189using the Exponential technique. The average percentage errors were determined and found to be 2.533884781using the linear technique. Comparing the three techniques results of the average percentage error, where it is found that the polynomial results have the minimum percentage average error, then the linear results and finally the exponential results. Therefore, the exponential technique is recommended for further research work. When the percentage error of the peak load squared, the sinusoidal form is becoming as a result for both the polynomial and the exponential, where the linear will take the shape of semi-sinusoidal form. In this case, the shape of the sinusoidal, which has been taken, will be homogeneous. Out of the results highlighted, it is recommended to use the polynomial, because it has less percentage error (better accuracy) and a sinusoidal form, which is homogeneous.

TABLE 10.12 The Estimated Peak Loads for the Kingdom of Bahrain with Calculated Results Using the Exponential

Year	Actual Peak Load (MW)	Estimated Peak Load (MW)	% Error (Peak Load)
2003	1535	1503.0584	2.080885993
2004	1632	1657.2612	–1.547867647
2005	1725	1811.464	–5.012405797
2006	1906	1965.6668	–3.130472193
2007	2136	2119.8696	0.755168539
2008	2314	2274.0724	1.725479689
2009	2438	2428.2752	0.398884331
2010	2708	2582.478	4.635228951
2011	2871	2736.6808	4.678481365
2012	2880	2890.8836	–0.377902778
2013	2917	3045.0864	–4.391031882
2014	3152	3199.2892	–1.500291878
2015	3441	3353.492	2.543097937
2016	3415	3507.6948	–2.714342606
2017	3572	3661.8976	–2.516730123
2018		3816.1004	
2019		3970.3032	
2020		4124.506	
2021		4278.7088	
2022		4432.9116	
2023		4587.1144	
2024		4741.3172	
2025		4895.52	

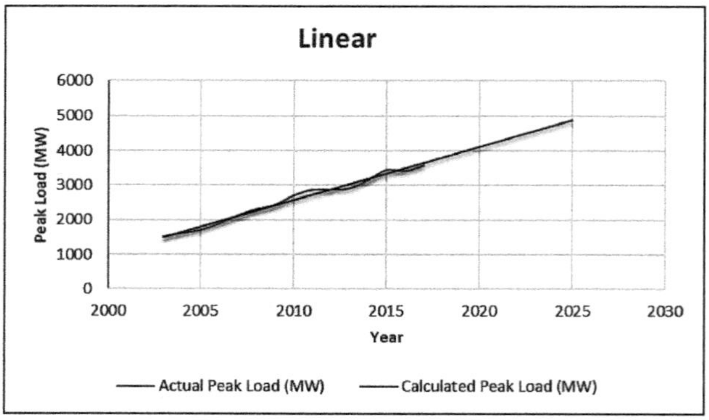

FIGURE 10.27 The estimated load versus the years.

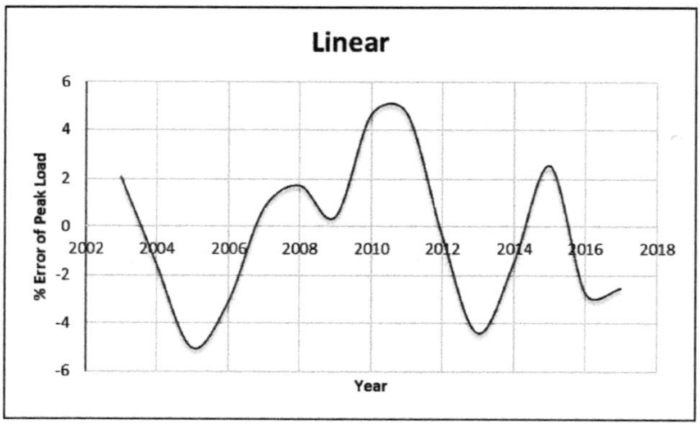

FIGURE 10.28 The percentage error for the load versus the years.

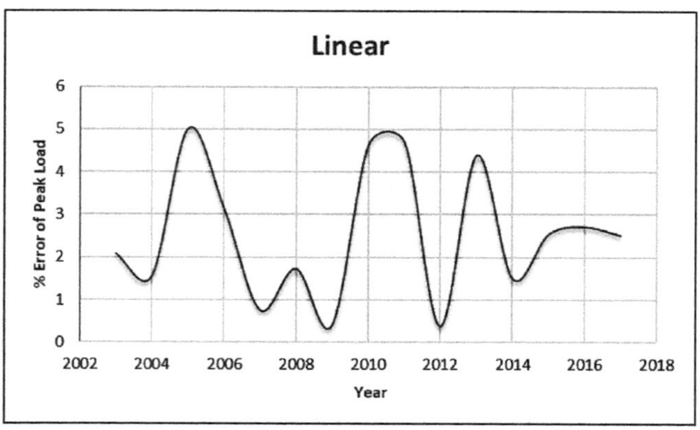

FIGURE 10.29 The absolute value of percentage error for the load versus the years.

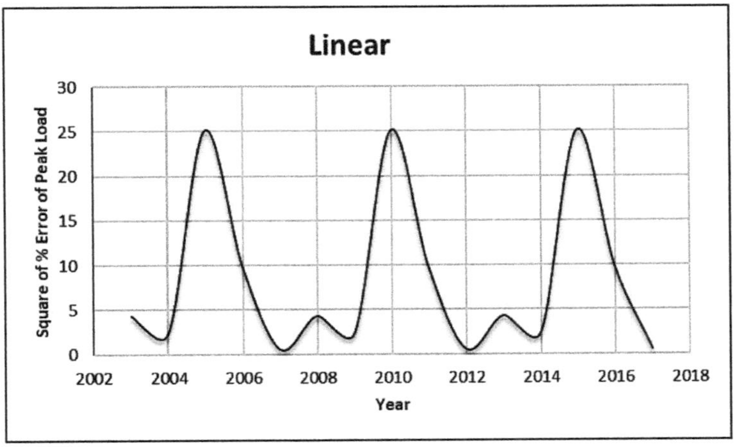

FIGURE 10.30 The square of the percentage error for the load versus the years.

KEYWORDS

- **adaptive network-based fuzzy inference system**
- **adaptive neuro-fuzzy inference system**
- **equivalent transition rate matrix**
- **load forecasted surface**
- **membership function**
- **three-state models**

REFERENCES

Ameen, I., & Novati, P. The solution of fractional order epidemic model by implicit Adams methods. *Applied Mathematical Model*, **2017,** *43*, 78–84.

Billinton, R., & Allan, R. N. *Reliability Evaluation of Engineering Systems: Concepts and Techniques* (2nd edn.). Pitman: New York, **1992**.

Qamber, I. S., & Al-Hamad, M. Y. Trading opportunities forecasted benefits at peak load for GCC countries. *Electric Power and Water Desalination Conference*, **2016,** Doha, Qatar.

Yanga, X., & Ralescu, D. Adams method for solving uncertain differential equations. *Applied Mathematics and Computation*, **2015,** *270*, 993–1003.

APPENDIX A

Selected Laplace Transforms Functions

The selected Laplace Transforms table is given below (Billinton and Allan, 1992):

f(t)	***F(s)***
1	$\frac{1}{s}$
t	$\frac{1}{s^2}$
e^{-kt}	$\frac{1}{s+k}$
$\sin kt$	$\frac{k}{s^2+k^2}$
$\cos kt$	$\frac{s}{s^2+k^2}$
$\frac{1}{(n-1)t} t^{n-1}\, e^{-kx}$	$\frac{1}{(s+k)^n}$
$\frac{dy}{dt}$	$sF(s)-y(0)$

APPENDIX B

Normal Distribution Table

The normal distribution table below gives the area under the normal probability density function between the limits of Z, where Z between 0 and $(x-\mu)/\sigma$ (Billinton and Allan, 1992):

Z	0.00	0.01	0.02	0.03	0.04	0.05	0.06	0.07	0.08	0.09
0.0	0.0000	0.0040	0.0080	0.0120	0.0159	0.0199	0.0239	0.0279	0.0319	0.0359
0.1	0.0398	0.0438	0.0478	0.0517	0.0557	0.0596	0.0636	0.0675	0.0714	0.0753
0.2	0.0793	0.0832	0.0871	0.0909	0.0948	0.0987	0.1026	0.1064	0.1103	0.1141
0.3	0.1179	0.1217	0.1255	0.1293	0.1331	0.1368	0.1406	0.1443	0.1480	0.1517
0.4	0.1555	0.1591	0.1628	0.1664	0.1700	0.1736	0.1772	0.1808	0.1844	0.1879
0.5	0.1915	0.1950	0.1985	0.2019	0.2054	0.2088	0.2123	0.2157	0.2190	0.2224
0.6	0.2257	0.2291	0.2324	0.2356	0.2389	0.2421	0.2454	0.2486	0.2517	0.2549
0.7	0.2580	0.2611	0.2642	0.2673	0.2703	0.2734	0.2764	0.2793	0.2823	0.2852
0.8	0.2881	0.2910	0.2939	0.2967	0.2995	0.3023	0.3051	0.3078	0.3108	0.3133
0.9	0.3159	0.3186	0.3212	0.3238	0.3264	0.3289	0.3315	0.3340	0.3365	0.3389
1.0	0.3413	0.3437	0.3461	0.3485	0.3508	0.3531	0.3554	0.3577	0.3599	0.3621
1.1	0.3643	0.3665	0.3686	0.3708	0.3729	0.3749	0.3770	0.3790	0.3810	0.3830
1.2	0.3849	0.3869	0.3888	0.3906	0.3925	0.3943	0.3962	0.3980	0.3997	0.4015
1.3	0.4032	0.4049	0.4066	0.4082	0.4099	0.4115	0.4131	0.4147	0.4162	0.4177
1.4	0.4192	0.4207	0.4222	0.4236	0.4251	0.4265	0.4279	0.4292	0.4306	0.4319
1.5	0.4332	0.4345	0.4357	0.4370	0.4382	0.4394	0.4406	0.4418	0.4429	0.4441
1.6	0.4452	0.4463	0.4474	0.4484	0.4495	0.4505	0.4515	0.4525	0.4535	0.4545
1.7	0.4554	0.4564	0.4573	0.4582	0.4591	0.4599	0.4608	0.4616	0.4625	0.4633
1.8	0.4641	0.4648	0.4656	0.4664	0.4671	0.4678	0.4686	0.4693	0.4699	0.4706
1.9	0.4713	0.4719	0.4726	0.4732	0.4738	0.4744	0.4750	0.4756	0.4761	0.4767
2.0	0.4772	0.4778	0.4783	0.4788	0.4793	0.4798	0.4803	0.4808	0.4812	0.4817
2.1	0.4821	0.4826	0.4830	0.4834	0.4838	0.4842	0.4846	0.4850	0.4854	0.4857

Continued table

Z	0.00	0.01	0.02	0.03	0.04	0.05	0.06	0.07	0.08	0.09
2.2	0.4861	0.4864	0.4868	0.4871	0.4874	0.4878	0.4881	0.4884	0.4887	0.4890
2.3	0.4893	0.4896	0.4898	0.4901	0.4904	0.4906	0.4909	0.4911	0.4913	0.4916
2.4	0.4918	0.4920	0.4922	0.4924	0.4927	0.4929	0.4930	0.4932	0.4934	0.4936
2.5	0.4938	0.4940	0.4941	0.4943	0.4945	0.4946	0.4948	0.4949	0.4951	0.4952
2.6	0.4953	0.4955	0.4956	0.4957	0.4958	0.4960	0.4961	0.4962	0.4963	0.4964
2.7	0.4965	0.4966	0.4967	0.4968	0.4969	0.4970	0.4971	0.4972	0.4973	0.4974
2.8	0.4974	0.4975	0.4976	0.4977	0.4977	0.4978	0.4979	0.4979	0.4980	0.4981
2.9	0.4981	0.4982	0.4982	0.4983	0.8384	0.4984	0.4985	0.4985	0.4986	0.4986
3.0	0.4986	0.4987	0.4987	0.4988	0.4988	0.4989	0.4989	0.4989	0.4990	0.4990
3.1	0.4990	0.4991	0.4991	0.4991	0.4991	0.4992	0.4992	0.4992	0.4993	0.4993
3.2	0.4993	0.4993	0.4994	0.4994	0.4994	0.4994	0.4994	0.4995	0.4995	0.4995
3.3	0.4995	0.4995	0.4995	0.4996	0.4996	0.4996	0.4996	0.4996	0.4996	0.4996
3.4	0.4997	0.4997	0.4997	0.4997	0.4997	0.4997	0.4997	0.4997	0.4997	0.4998

APPENDIX C

LOLP Calculation Steps

Assuming that the average peak loads are distributed over one year in days, as shown in Figure C.1.

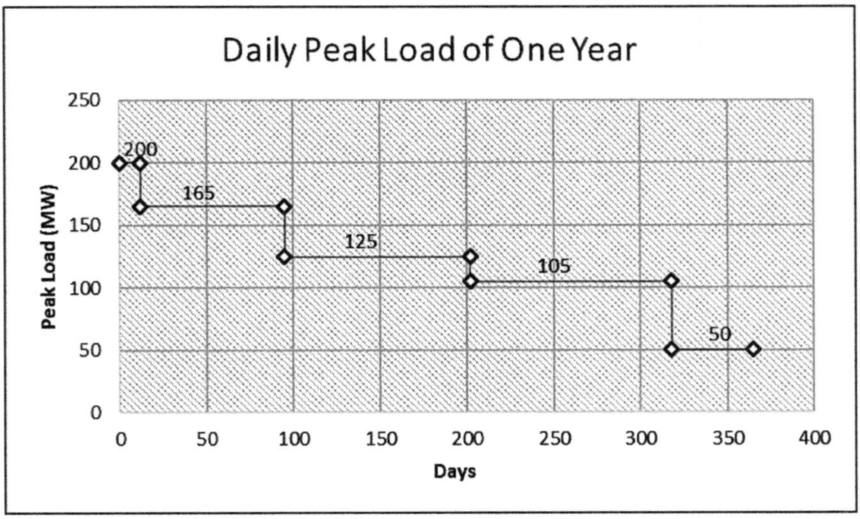

FIGURE C.1 Average peak loads distribution over one year.

Calculating the LOLP as illustrated in Table C.1.

TABLE C.1 Calculation of LOLP

Capacity IN (MW)	Capacity OUT (MW)	$P_i t_i$
200	80	3.20×10^{-4}
165	115	2.38×10^{-4}
125	155	6.69×10^{-6}
105	175	4.42×10^{-6}
50	230	6.06×10^{-9}

Therefore, the LOLP is equal to 5.69×10^{-4}.

Figure C.2 illustrates the distribution of each average peak load.

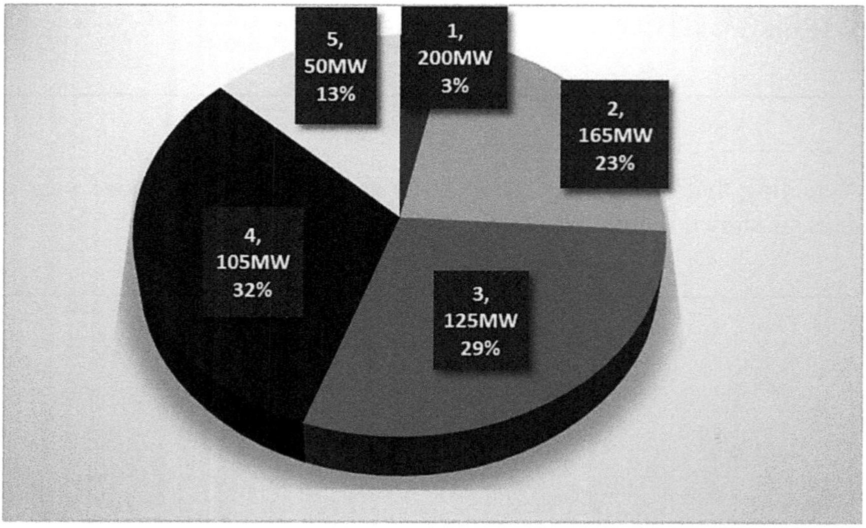

FIGURE C.2 The distribution of each average peak load.

Index

M

N

O

P

R

S

T

U

V

W

Z